M000237085

WORLD FLAGS
Color By Number
For Adults

THIS COLORING BOOK BELONGS TO:

_____

_____

B.C. Lester Books
Geography publications for the people since 2019.

Visit us at www.bclesterbooks.com for more!

A MESSAGE FROM THE PUBLISHER

Hey! Thank you for making the purchase, we really hope you enjoy this book. If you have the chance, then all feedback is greatly appreciated. We have put a lot of effort into making this book, so if you are not completely satisfied, please email us at ben@bclesterbooks.com and we will do our best to address the issues. If you have any suggestions, enquries or want to send us a selfie with this book, then email at the same address – ben@bclesterbooks.com

Is this book misprinted? Drop us an email with a photo of the misprint and we will send out another copy!

WHO ARE WE AT B.C. LESTER BOOKS?

B.C. Lester Books is a small publishing firm of three people based in Buckinghamshire, UK. We aim to provide quality works in all things geography, for kids and adults, with varying interests. We have already released a selection of activity, trivia and fact books and are working hard to bring you wider selection. Have a suggestion for us? Then email ben@bclesterbooks.com. We are all ears!

HAVE FUN WITH OUR GIFT TO YOU: A 3-IN-1 GEOGRAPHY QUIZ BOOK!

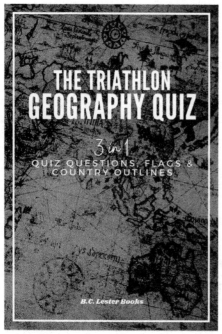

Go here to grab your FREE copy!
www.bclesterbooks.com/freebies/

CONTENTS

BEFORE YOU START...

Test your coloring equipment here for bleedthrough. Note that this coloring book is **NOT** recommended for paint or highlighters...

HOW TO USE THIS BOOK

Color By Number

On each flag presented, numbers will appear. Each number corresponds to a color shown in the key displayed by it. To use, simply color in the spaces according to the key. **Note: if there is no number assigned to a space, then leave it as white!**

(flag with stars marked 14)	2
	2
	2
	2
	2
	2
	2

2 = DARK RED
14= INDIGO

Not all colors available?

If only limited colors are available, combine *1 and 2 (**dark red**) - 4 and 5 (**orange**) - 6 and 7 (**yellow**) - 8 and 9 (**green**) - 12 and 13 (**blue**)*. Colors *15, 17 and 18 (purple, grey, brown)* only occur in tiny amounts across the world flags, so can be ignored.

Best coloring equipment to use:

The nature of country flags and their varying detail (see *Japan* and *Bhutan* below!) means that **coloring pencils** will be the best and most precise for you to use and what this book has been designed for. If you choose to use other equipment, be sure to check whether it is suitable for use in this book on the previous page!

JAPAN

BHUTAN

Accuracy of flags portrayed:

Since over 500 colors are needed for complete accuracy, some colors have been grouped into the closest of the given 18 colors. Some of the smallest flag coloring details have also been ignored by the numbering system for ease of use, but in all cases, the major details and structure have been left untouched!

ALL GOOD?

Prepare your favorite brew, relax, and **enjoy** the experience!

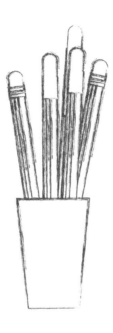

NORTH AMERICAN COUNTRY FLAGS

ANTIGUA & BARBUDA

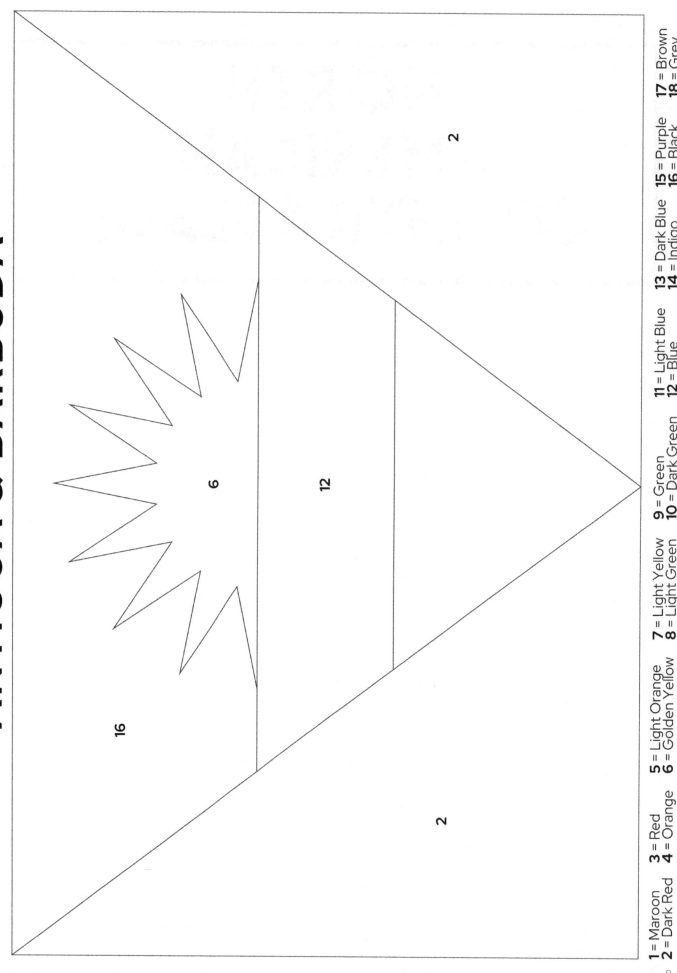

1 = Maroon
2 = Dark Red

3 = Red
4 = Orange

5 = Light Orange
6 = Golden Yellow

7 = Light Yellow
8 = Light Green

9 = Green
10 = Dark Green

11 = Light Blue
12 = Blue

13 = Dark Blue
14 = Indigo

15 = Purple
16 = Black

17 = Brown
18 = Grey

THE BAHAMAS

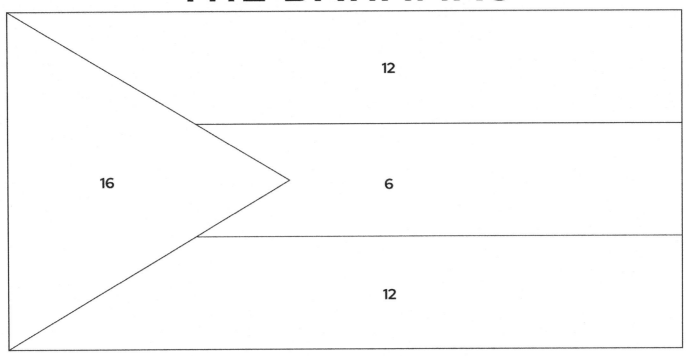

BARBADOS

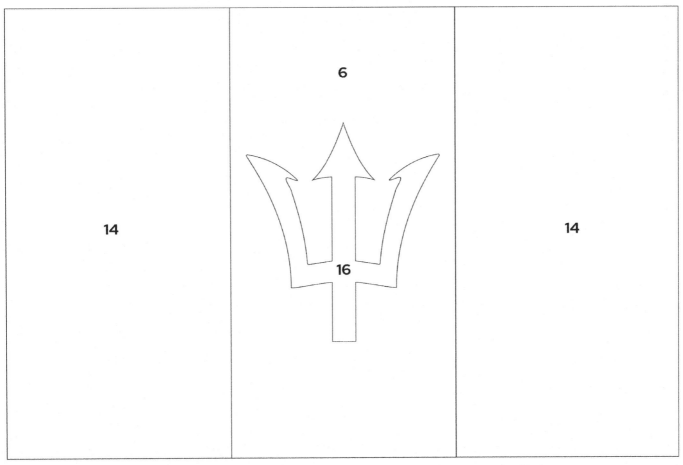

1 = Maroon	**4** = Orange	**7** = Light Yellow	**10** = Dark Green	**13** = Dark Blue	**16** = Black
2 = Dark Red	**5** = Light Orange	**8** = Light Green	**11** = Light Blue	**14** = Indigo	**17** = Grey
3 = Red	**6** = Golden Yellow	**9** = Green	**12** = Blue	**15** = Purple	**18** = Brown

BELIZE

2

13

SUB UMBRA FLOREO

13

2

CANADA

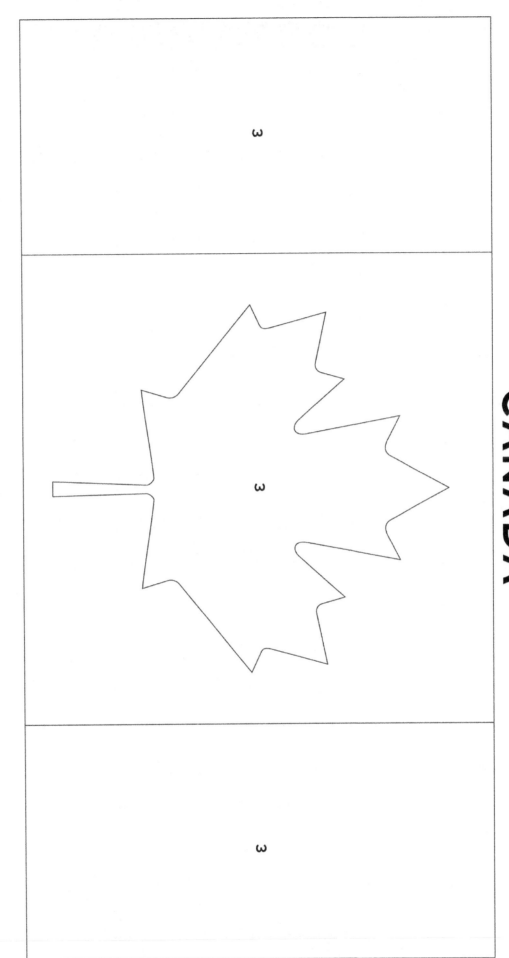

3

3

3

1 = Maroon
2 = Dark Red

3 = Red
4 = Orange

5 = Light Orange
6 = Golden Yellow

7 = Light Yellow
8 = Light Green

9 = Green
10 = Dark Green

11 = Light Blue
12 = Blue

13 = Dark Blue
14 = Indigo

15 = Purple
16 = Black

17 = Brown
18 = Grey

COSTA RICA

CUBA

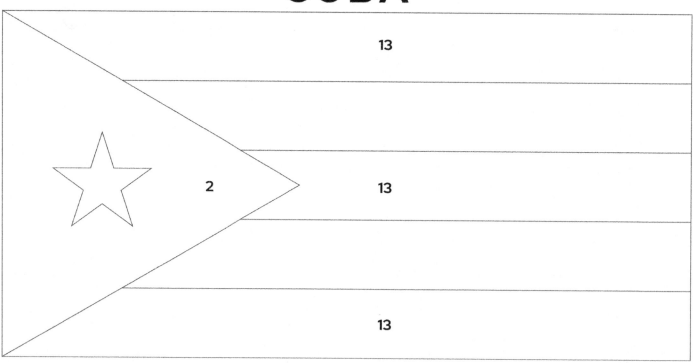

1 = Maroon   4 = Orange   7 = Light Yellow   10 = Dark Green   13 = Dark Blue   16 = Black
2 = Dark Red   5 = Light Orange   8 = Light Green   11 = Light Blue   14 = Indigo   17 = Grey
3 = Red   6 = Golden Yellow 9 = Green   12 = Blue   15 = Purple   18 = Brown

DOMINICA

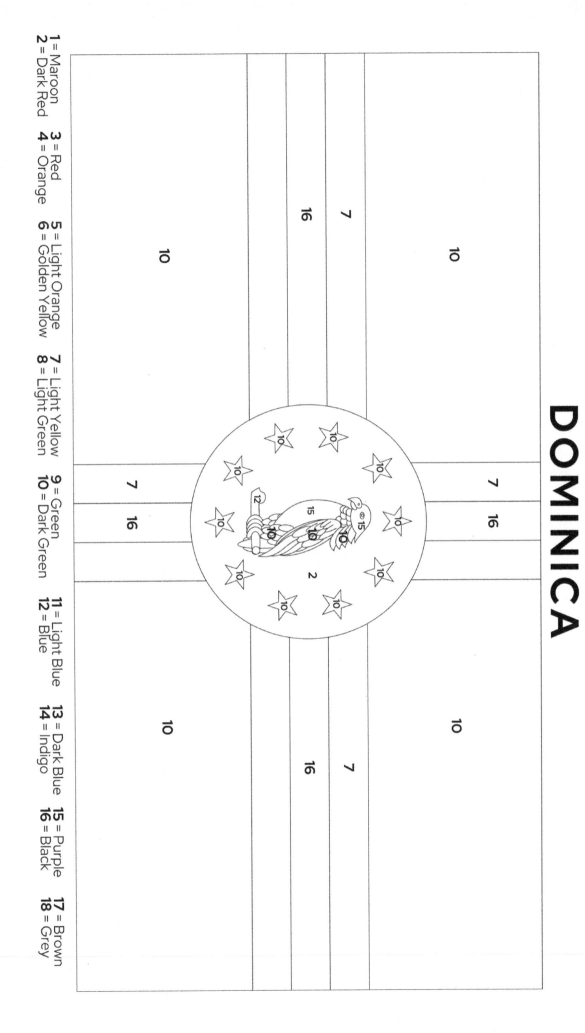

1 = Maroon
2 = Dark Red
3 = Red
4 = Orange
5 = Light Orange
6 = Golden Yellow
7 = Light Yellow
8 = Light Green
9 = Green
10 = Dark Green
11 = Light Blue
12 = Blue
13 = Dark Blue
14 = Indigo
15 = Purple
16 = Black
17 = Brown
18 = Grey

DOMINICAN REPUBLIC

14

2

2

14

1 = Maroon
2 = Dark Red

3 = Red
4 = Orange

5 = Light Orange
6 = Golden Yellow

7 = Light Yellow
8 = Light Green

9 = Green
10 = Dark Green

11 = Light Blue
12 = Blue

13 = Dark Blue
14 = Indigo

15 = Purple
16 = Black

17 = Brown
18 = Grey

EL SALVADOR

13

13

13

REPUBLICA DE EL SALVADOR EN LA AMERICA CENTRAL

13
13
13
13

1 = Maroon
2 = Dark Red
3 = Red
4 = Orange
5 = Light Orange
6 = Golden Yellow
7 = Light Yellow
8 = Light Green
9 = Green
10 = Dark Green
11 = Light Blue
12 = Blue
13 = Dark Blue
14 = Indigo
15 = Purple
16 = Black
17 = Brown
18 = Grey

GRENADA

HAITI

13

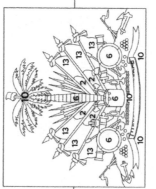

2

1 = Maroon   3 = Red   5 = Light Orange   7 = Light Yellow   9 = Green   11 = Light Blue   13 = Dark Blue   15 = Purple   17 = Brown
2 = Dark Red   4 = Orange   6 = Golden Yellow   8 = Light Green   10 = Dark Green   12 = Blue   14 = Indigo   16 = Black   18 = Grey

HONDURAS

1 = Maroon
2 = Dark Red

3 = Red
4 = Orange

5 = Light Orange
6 = Golden Yellow

7 = Light Yellow
8 = Light Green

9 = Green
10 = Dark Green

11 = Light Blue
12 = Blue

13 = Dark Blue
14 = Indigo

15 = Purple
16 = Black

17 = Brown
18 = Grey

13

13

13

13

13

13

13

13

JAMAICA

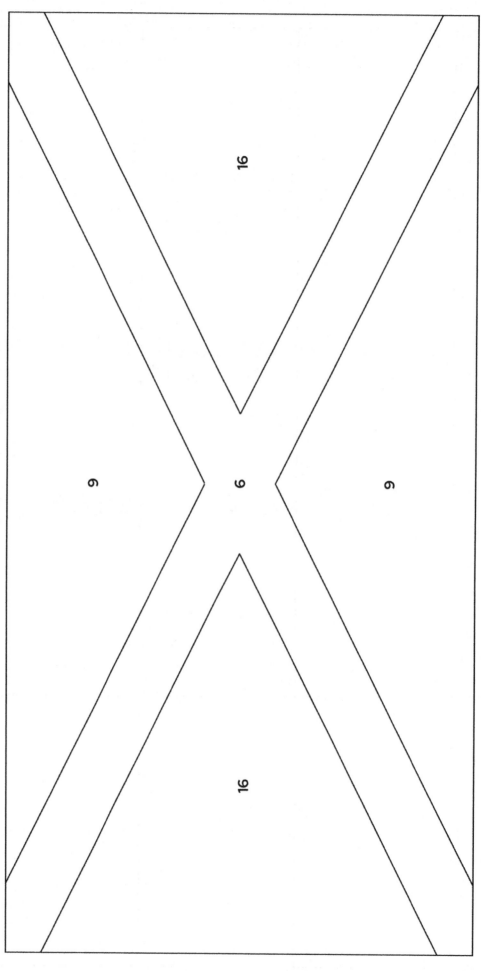

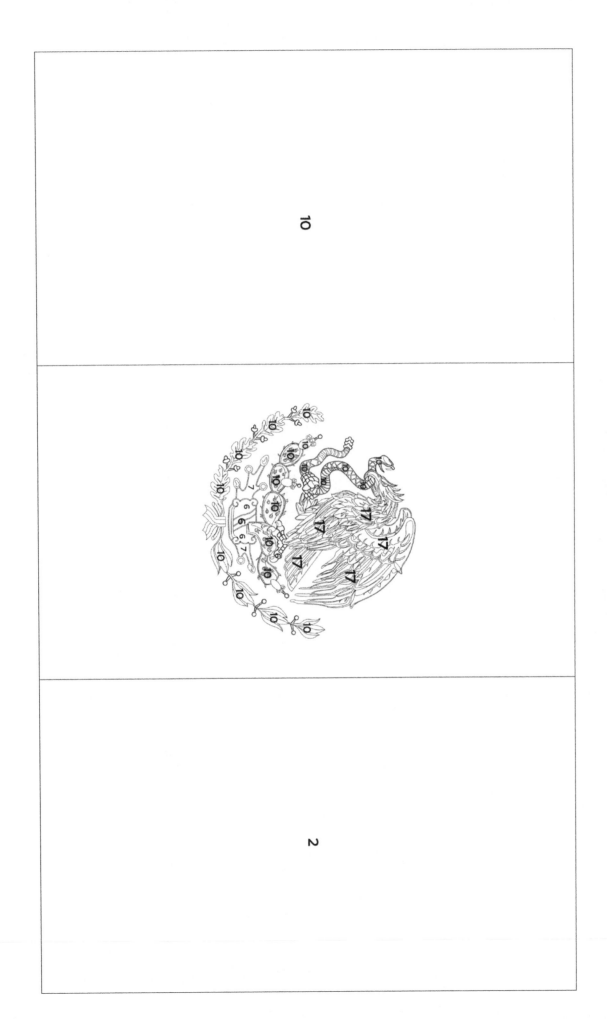

1 = Maroon
2 = Dark Red
3 = Red
4 = Orange
5 = Light Orange
6 = Golden Yellow
7 = Light Yellow
8 = Light Green
9 = Green
10 = Dark Green
11 = Light Blue
12 = Blue
13 = Dark Blue
14 = Indigo
15 = Purple
16 = Black
17 = Brown
18 = Grey

NICARAGUA

12

12

REPUBLICA DE NICARAGUA · AMERICA CENTRAL

1 = Maroon   **3** = Red   **5** = Light Orange   **7** = Light Yellow   **9** = Green   **11** = Light Blue   **13** = Dark Blue   **15** = Purple   **17** = Brown
2 = Dark Red   **4** = Orange   **6** = Golden Yellow   **8** = Light Green   **10** = Dark Green   **12** = Blue   **14** = Indigo   **16** = Black   **18** = Grey

PANAMA

1 = Maroon
2 = Dark Red
3 = Red
4 = Orange
5 = Light Orange
6 = Golden Yellow
7 = Light Yellow
8 = Light Green
9 = Green
10 = Dark Green
11 = Light Blue
12 = Blue
13 = Dark Blue
14 = Indigo
15 = Purple
16 = Black
17 = Brown
18 = Grey

14

14

2

2

SAINT KITTS AND NEVIS

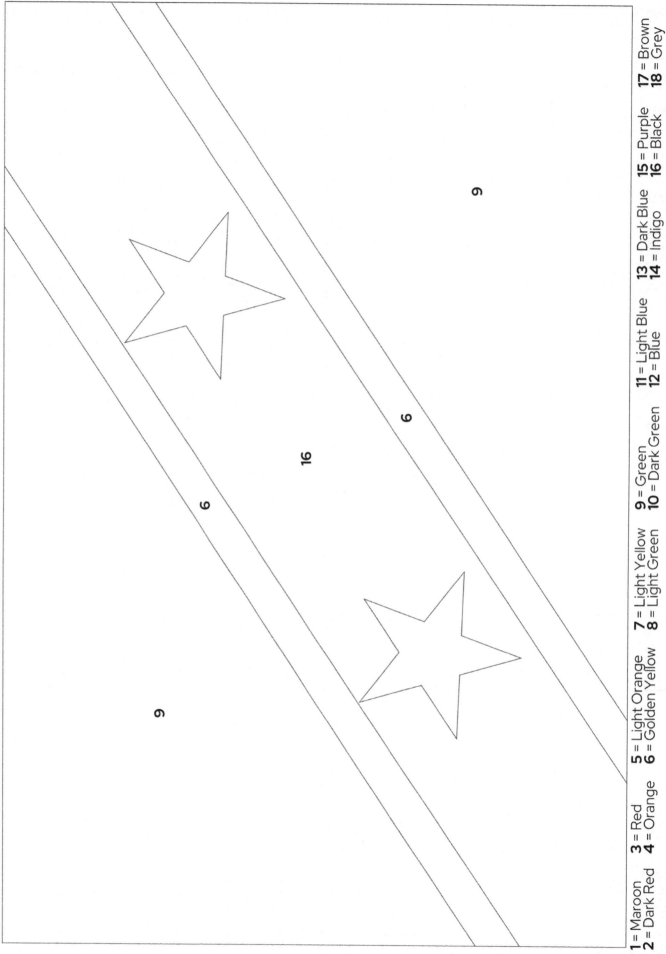

SAINT LUCIA

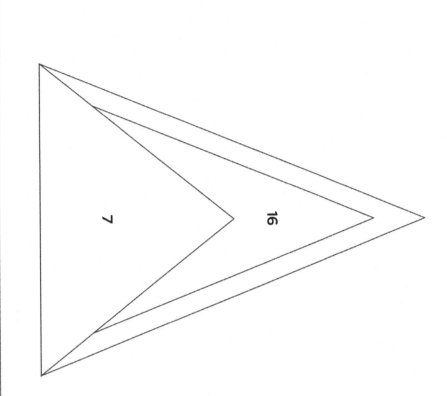

1 = Maroon
2 = Dark Red
3 = Red
4 = Orange
5 = Light Orange
6 = Golden Yellow
7 = Light Yellow
8 = Light Green
9 = Green
10 = Dark Green
11 = Light Blue
12 = Blue
13 = Dark Blue
14 = Indigo
15 = Purple
16 = Black
17 = Brown
18 = Grey

SAINT VINCENT AND THE GRENADINES

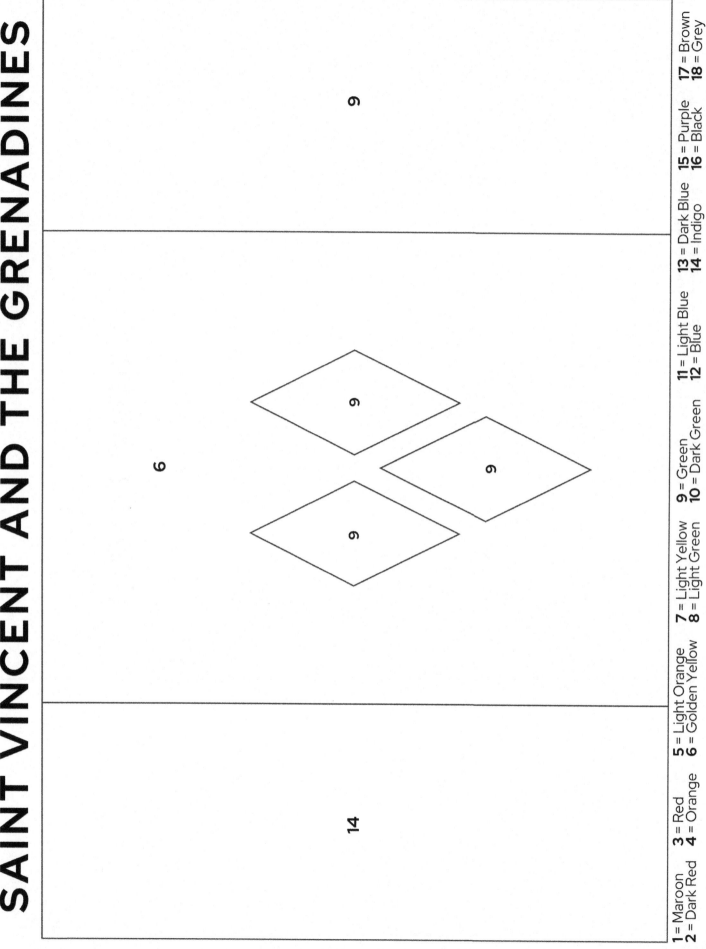

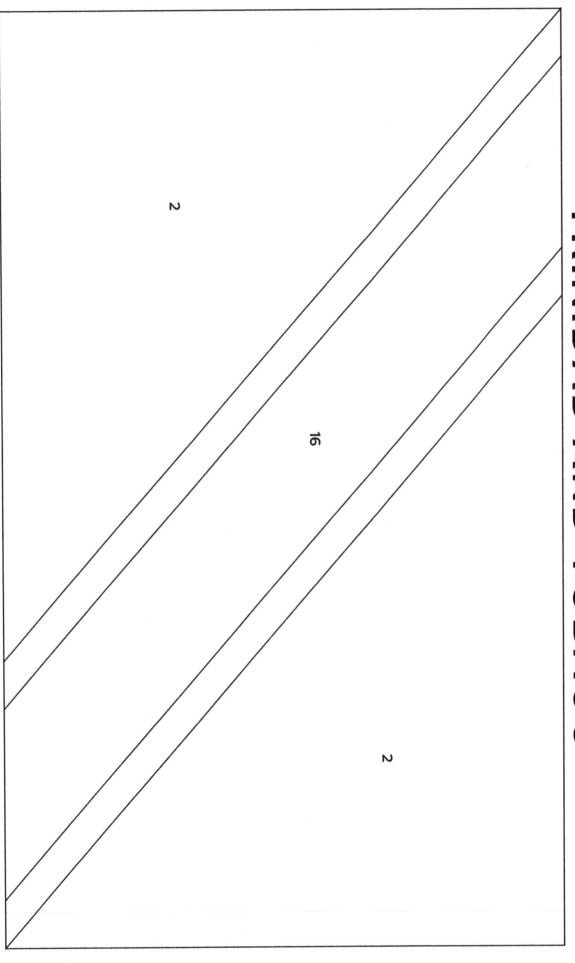

TRINIDAD AND TOBAGO

1 = Maroon
2 = Dark Red

3 = Red
4 = Orange

5 = Light Orange
6 = Golden Yellow

7 = Light Yellow
8 = Light Green

9 = Green
10 = Dark Green

11 = Light Blue
12 = Blue

13 = Dark Blue
14 = Indigo

15 = Purple
16 = Black

17 = Brown
18 = Grey

UNITED STATES OF AMERICA

2

2

2

2

2

2

2

14

14

1 = Maroon
2 = Dark Red

3 = Red
4 = Orange

5 = Light Orange
6 = Golden Yellow

7 = Light Yellow
8 = Light Green

9 = Green
10 = Dark Green

11 = Light Blue
12 = Blue

13 = Dark Blue
14 = Indigo

15 = Purple
16 = Black

17 = Brown
18 = Grey

SOUTH AMERICAN COUNTRY FLAGS

ARGENTINA

11

11

BOLIVIA

3	6	9

1 = Maroon **3** = Red **5** = Light Orange **7** = Light Yellow **9** = Green **11** = Light Blue **13** = Dark Blue **15** = Purple **17** = Brown
2 = Dark Red **4** = Orange **6** = Golden Yellow **8** = Light Green **10** = Dark Green **12** = Blue **14** = Indigo **16** = Black **18** = Grey

BRAZIL

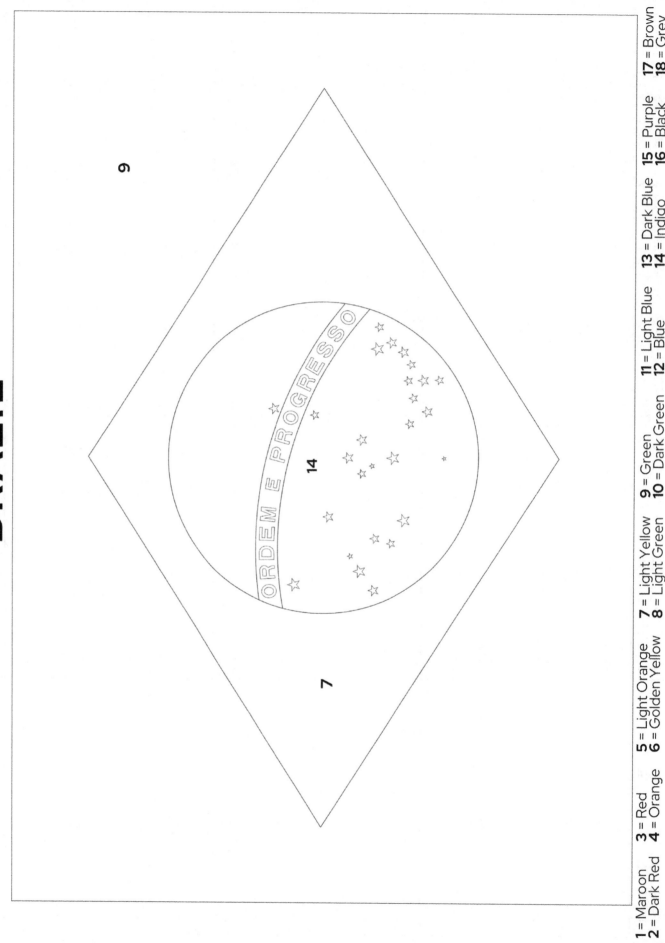

ORDEM E PROGRESSO

9

7

14

CHILE

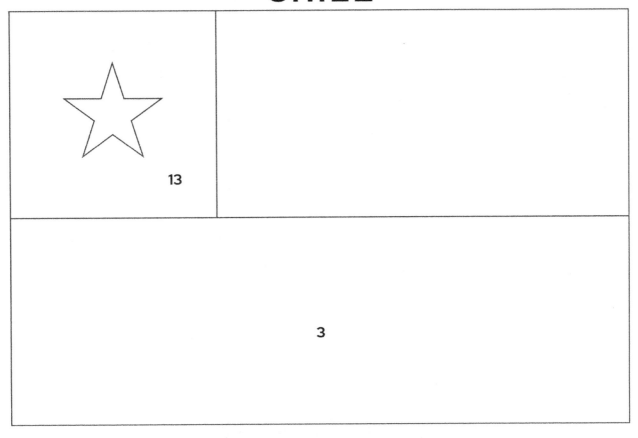

13

3

COLOMBIA

6

13

2

ECUADOR

1 = Maroon	**3** = Red	**5** = Light Orange	**7** = Light Yellow	**9** = Green	**11** = Light Blue	**13** = Dark Blue	**15** = Purple	**17** = Brown
2 = Dark Red	**4** = Orange	**6** = Golden Yellow	**8** = Light Green	**10** = Dark Green	**12** = Blue	**14** = Indigo	**16** = Black	**18** = Grey

GUYANA

1 = Maroon
2 = Dark Red
3 = Red
4 = Orange
5 = Light Orange
6 = Golden Yellow
7 = Light Yellow
8 = Light Green
9 = Green
10 = Dark Green
11 = Light Blue
12 = Blue
13 = Dark Blue
14 = Indigo
15 = Purple
16 = Black
17 = Brown
18 = Grey

2

16

7

9

9

PARAGUAY

2

13

REPÚBLICA DEL PARAGUAY

1 = Maroon   3 = Red   5 = Light Orange   7 = Light Yellow   9 = Green   11 = Light Blue   13 = Dark Blue   15 = Purple   17 = Brown
2 = Dark Red   4 = Orange   6 = Golden Yellow   8 = Light Green   10 = Dark Green   12 = Blue   14 = Indigo   16 = Black   18 = Grey

PERU

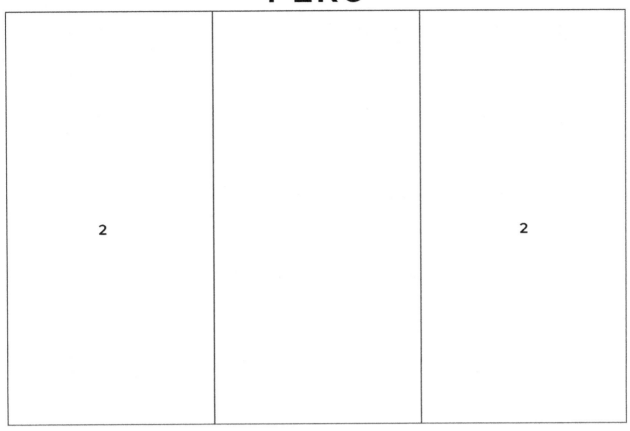

SURINAME

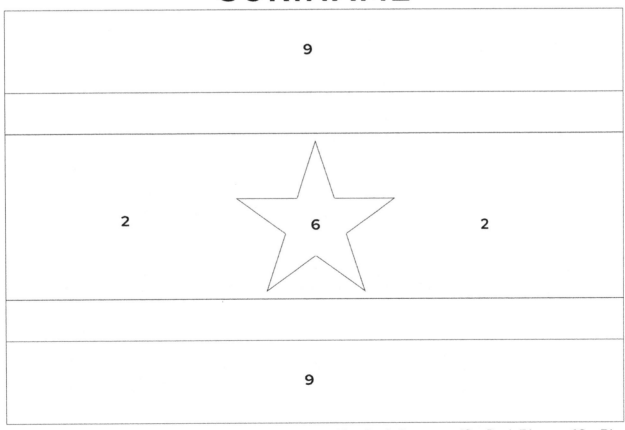

1 = Maroon   4 = Orange   7 = Light Yellow   10 = Dark Green   13 = Dark Blue   16 = Black
2 = Dark Red   5 = Light Orange   8 = Light Green   11 = Light Blue   14 = Indigo   17 = Grey
3 = Red   6 = Golden Yellow 9 = Green   12 = Blue   15 = Purple   18 = Brown

URUGUAY

13

13

13

13

VENEZUELA

6

14

2

EUROPEAN COUNTRY FLAGS

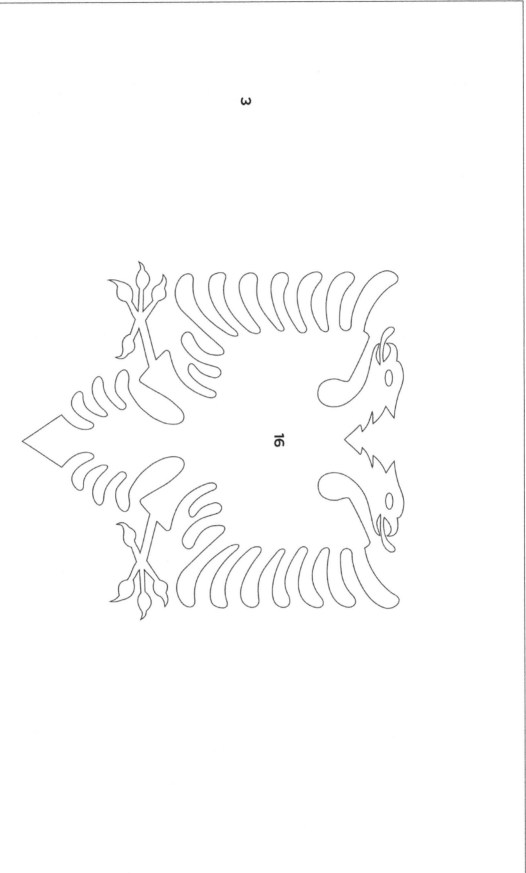

ANDORRA

VIRTVS VNITA FORTIOR

2

13

AUSTRIA

3	3	3

1 = Maroon   **3** = Red   **5** = Light Orange   **7** = Light Yellow   **9** = Green   **11** = Light Blue   **13** = Dark Blue   **15** = Purple   **17** = Brown
2 = Dark Red   **4** = Orange   **6** = Golden Yellow   **8** = Light Green   **10** = Dark Green   **12** = Blue   **14** = Indigo   **16** = Black   **18** = Grey

BELARUS

2

10

BELGIUM

16

6

3

1 = Maroon   4 = Orange   7 = Light Yellow   10 = Dark Green   13 = Dark Blue   16 = Black
2 = Dark Red   5 = Light Orange   8 = Light Green   11 = Light Blue   14 = Indigo   17 = Grey
3 = Red   6 = Golden Yellow   9 = Green   12 = Blue   15 = Purple   18 = Brown

BOSNIA AND HERZEGOVINA

14

3

14

1 = Maroon
2 = Dark Red

3 = Red
4 = Orange

5 = Light Orange
6 = Golden Yellow

7 = Light Yellow
8 = Light Green

9 = Green
10 = Dark Green

11 = Light Blue
12 = Blue

13 = Dark Blue
14 = Indigo

15 = Purple
16 = Black

17 = Brown
18 = Grey

BULGARIA

1 = Maroon
2 = Dark Red
3 = Red
4 = Orange
5 = Light Orange
6 = Golden Yellow
7 = Light Yellow
8 = Light Green
9 = Green
10 = Dark Green
11 = Light Blue
12 = Blue
13 = Dark Blue
14 = Indigo
15 = Purple
16 = Black
17 = Brown
18 = Grey

2

9

CROATIA

2

2

14

2

2   2
   2
2   2
   2   2
2   2
   2
2   2
      2

14

CYPRUS

1 = Maroon
2 = Dark Red

3 = Red
4 = Orange

5 = Light Orange
6 = Golden Yellow

7 = Light Yellow
8 = Light Green

9 = Green
10 = Dark Green

11 = Light Blue
12 = Blue

13 = Dark Blue
14 = Indigo

15 = Purple
16 = Black

17 = Brown
18 = Grey

CZECH REPUBLIC

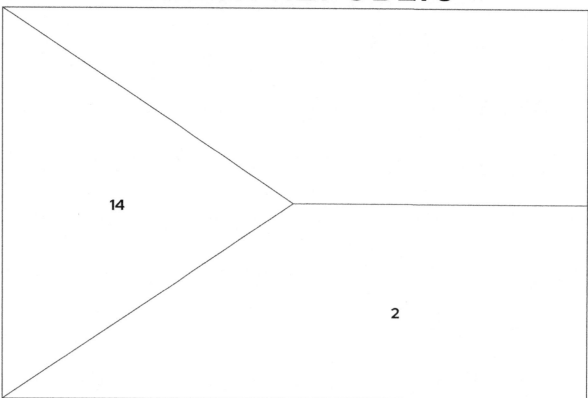

DENMARK

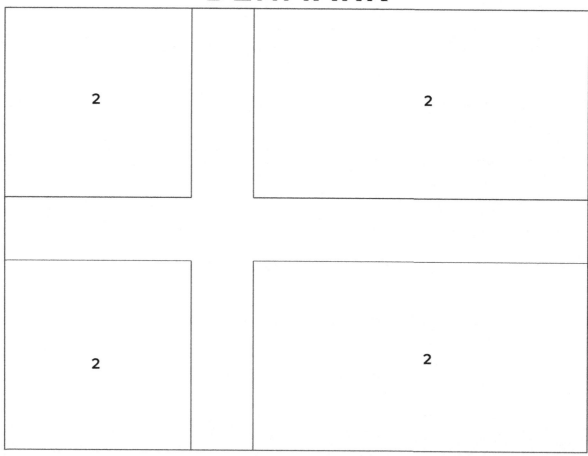

1 = Maroon 4 = Orange 7 = Light Yellow 10 = Dark Green 13 = Dark Blue 16 = Black
2 = Dark Red 5 = Light Orange 8 = Light Green 11 = Light Blue 14 = Indigo 17 = Grey
3 = Red 6 = Golden Yellow 9 = Green 12 = Blue 15 = Purple 18 = Brown

ESTONIA

12
16

FINLAND

14

FRANCE

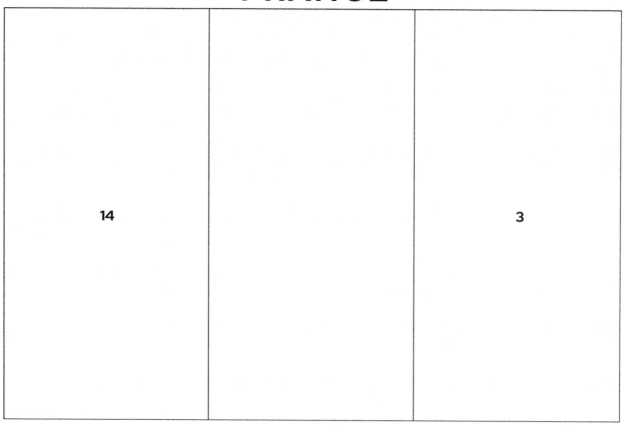

14		3

GERMANY

16
3
6

1 = Maroon   **4** = Orange   **7** = Light Yellow   **10** = Dark Green   **13** = Dark Blue   **16** = Black
2 = Dark Red   **5** = Light Orange   **8** = Light Green   **11** = Light Blue   **14** = Indigo   **17** = Grey
3 = Red   **6** = Golden Yellow   **9** = Green   **12** = Blue   **15** = Purple   **18** = Brown

GREECE

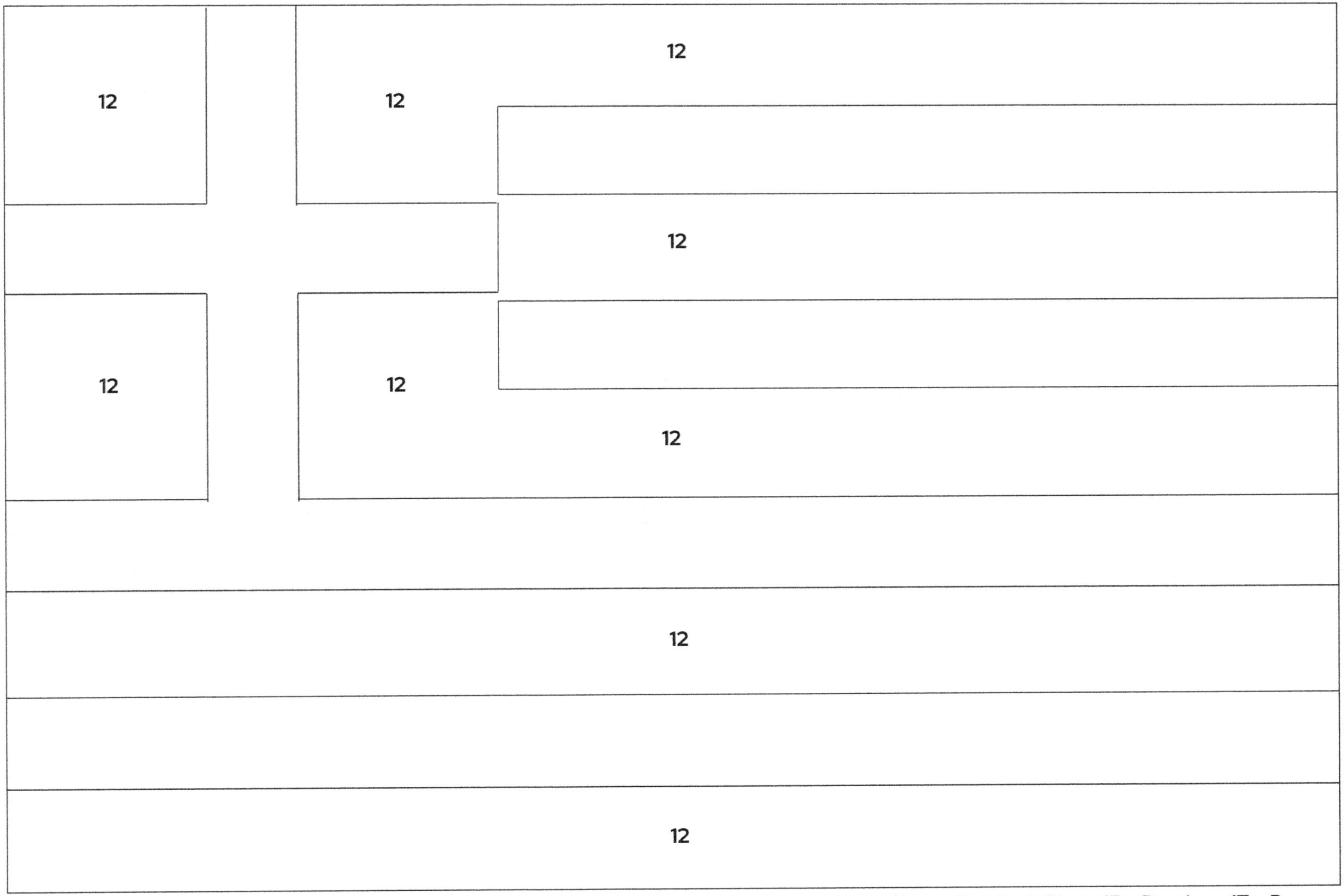
12
12
12
12
12
12
12
12
12

1 = Maroon **2** = Dark Red **3** = Red **4** = Orange **5** = Light Orange **6** = Golden Yellow **7** = Light Yellow **8** = Light Green **9** = Green **10** = Dark Green **11** = Light Blue **12** = Blue **13** = Dark Blue **14** = Indigo **15** = Purple **16** = Black **17** = Brown **18** = Grey

HUNGARY

2
10

ICELAND

IRELAND

9		5

ITALY

9		2

1 = Maroon	**4** = Orange	**7** = Light Yellow	**10** = Dark Green	**13** = Dark Blue	**16** = Black
2 = Dark Red	**5** = Light Orange	**8** = Light Green	**11** = Light Blue	**14** = Indigo	**17** = Grey
3 = Red	**6** = Golden Yellow	**9** = Green	**12** = Blue	**15** = Purple	**18** = Brown

KOSOVO

13

6

LATVIA

	1	1

1 = Maroon
2 = Dark Red

3 = Red
4 = Orange

5 = Light Orange
6 = Golden Yellow

7 = Light Yellow
8 = Light Green

9 = Green
10 = Dark Green

11 = Light Blue
12 = Blue

13 = Dark Blue
14 = Indigo

15 = Purple
16 = Black

17 = Brown
18 = Grey

LIECHTENSTEIN

14

2

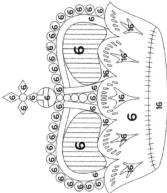

1 = Maroon
2 = Dark Red
3 = Red
4 = Orange
5 = Light Orange
6 = Golden Yellow
7 = Light Yellow
8 = Light Green
9 = Green
10 = Dark Green
11 = Light Blue
12 = Blue
13 = Dark Blue
14 = Indigo
15 = Purple
16 = Black
17 = Brown
18 = Grey

LITHUANIA

6
10
2

LUXEMBOURG

3
11

MALTA

2

18

18   18

FOR GALLANTRY
18 17 18
18   18

1 = Maroon   3 = Red   5 = Light Orange   7 = Light Yellow   9 = Green   11 = Light Blue   13 = Dark Blue   15 = Purple   17 = Brown
2 = Dark Red   4 = Orange   6 = Golden Yellow   8 = Light Green   10 = Dark Green   12 = Blue   14 = Indigo   16 = Black   18 = Grey

1 = Maroon
2 = Dark Red
3 = Red
4 = Orange
5 = Light Orange
6 = Golden Yellow
7 = Light Yellow
8 = Light Green
9 = Green
10 = Dark Green
11 = Light Blue
12 = Blue
13 = Dark Blue
14 = Indigo
15 = Purple
16 = Black
17 = Brown
18 = Grey

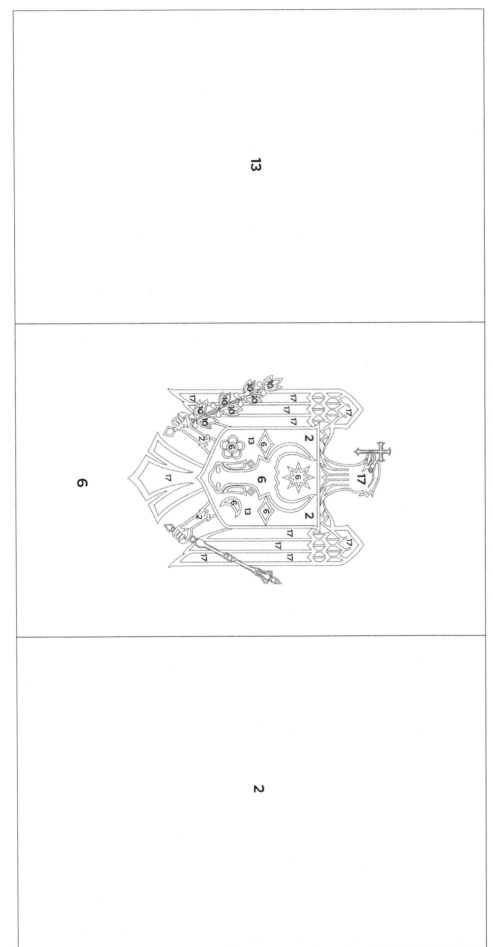

MONACO

3

1 = Maroon   4 = Orange   7 = Light Yellow   10 = Dark Green   13 = Dark Blue   16 = Black
2 = Dark Red   5 = Light Orange   8 = Light Green   11 = Light Blue   14 = Indigo   17 = Grey
3 = Red   6 = Golden Yellow   9 = Green   12 = Blue   15 = Purple   18 = Brown

MONTENEGRO

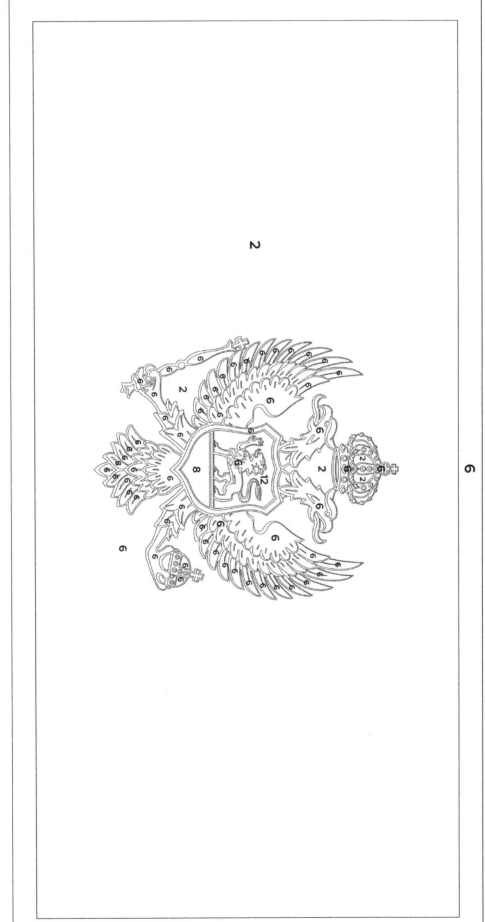

THE NETHERLANDS

2

13

1 = Maroon   3 = Red   5 = Light Orange   7 = Light Yellow   9 = Green   11 = Light Blue   13 = Dark Blue   15 = Purple   17 = Brown
2 = Dark Red 4 = Orange   6 = Golden Yellow   8 = Light Green   10 = Dark Green 12 = Blue   14 = Indigo   16 = Black   18 = Grey

NORTH MACEDONIA

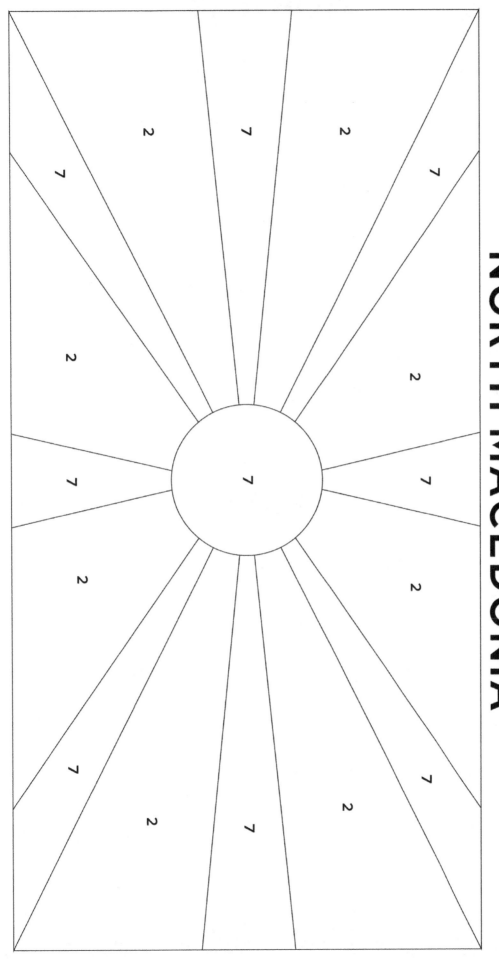

1 = Maroon
2 = Dark Red
3 = Red
4 = Orange
5 = Light Orange
6 = Golden Yellow
7 = Light Yellow
8 = Light Green
9 = Green
10 = Dark Green
11 = Light Blue
12 = Blue
13 = Dark Blue
14 = Indigo
15 = Purple
16 = Black
17 = Brown
18 = Grey

NORWAY

POLAND

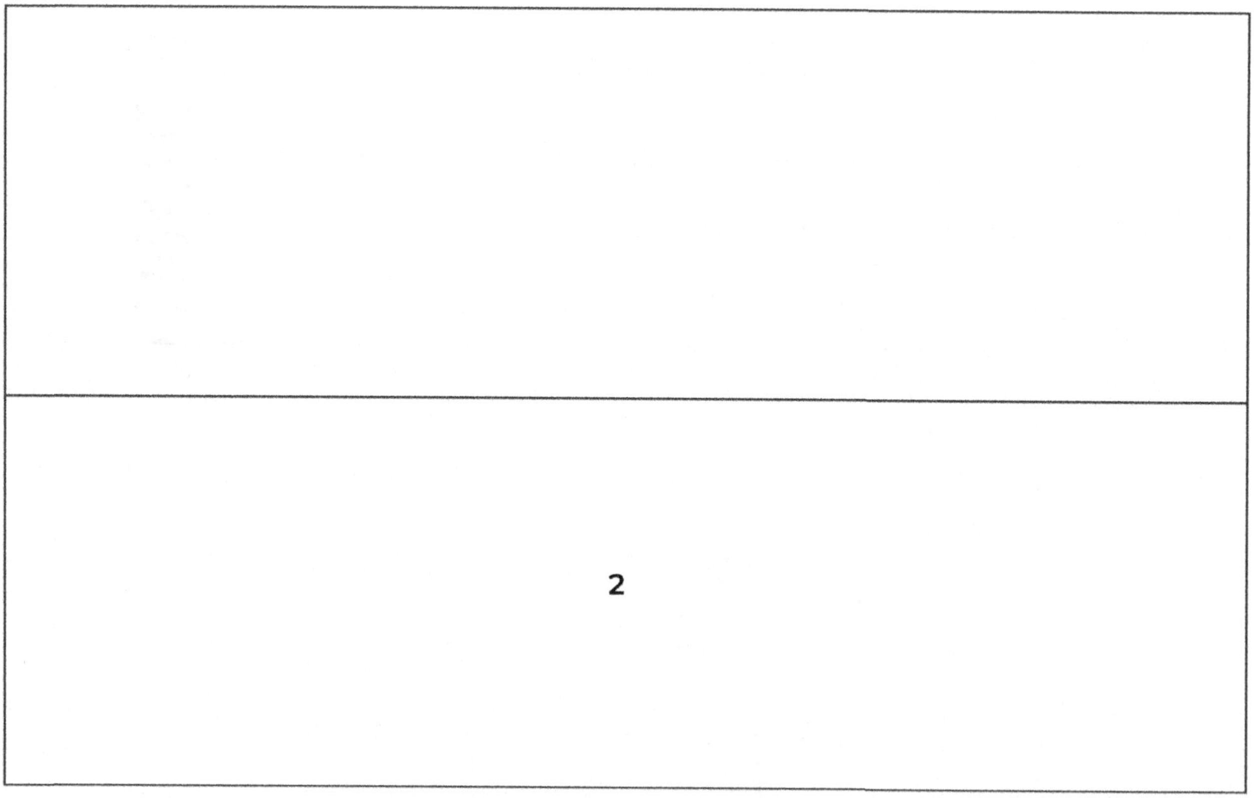

1 = Maroon
2 = Dark Red
3 = Red
4 = Orange
5 = Light Orange
6 = Golden Yellow
7 = Light Yellow
8 = Light Green
9 = Green
10 = Dark Green
11 = Light Blue
12 = Blue
13 = Dark Blue
14 = Indigo
15 = Purple
16 = Black
17 = Grey
18 = Brown

PORTUGAL

ROMANIA

14	6	2

RUSSIA

13
3

1 = Maroon   4 = Orange   7 = Light Yellow   10 = Dark Green   13 = Dark Blue   16 = Black
2 = Dark Red   5 = Light Orange   8 = Light Green   11 = Light Blue   14 = Indigo   17 = Grey
3 = Red   6 = Golden Yellow   9 = Green   12 = Blue   15 = Purple   18 = Brown

68

11

8

LIBERTAS

18 18 18 11 11

1 = Maroon **4** = Orange **7** = Light Yellow **10** = Dark Green **13** = Dark Blue **16** = Black
2 = Dark Red **5** = Light Orange **8** = Light Green **11** = Light Blue **14** = Indigo **17** = Grey
3 = Red **6** = Golden Yellow **9** = Green **12** = Blue **15** = Purple **18** = Brown

SERBIA

2

14

14

70

SLOVAKIA

1 = Maroon
2 = Dark Red
3 = Red
4 = Orange
5 = Light Orange
6 = Golden Yellow
7 = Light Yellow
8 = Light Green
9 = Green
10 = Dark Green
11 = Light Blue
12 = Blue
13 = Dark Blue
14 = Indigo
15 = Purple
16 = Black
17 = Brown
18 = Grey

13

13

13

3

3

3

3

SLOVENIA

13

3

1 = Maroon
2 = Dark Red

3 = Red
4 = Orange

5 = Light Orange
6 = Golden Yellow

7 = Light Yellow
8 = Light Green

9 = Green
10 = Dark Green

11 = Light Blue
12 = Blue

13 = Dark Blue
14 = Indigo

15 = Purple
16 = Black

17 = Brown
18 = Grey

1 = Maroon
2 = Dark Red
3 = Red
4 = Orange
5 = Light Orange
6 = Golden Yellow
7 = Light Yellow
8 = Light Green
9 = Green
10 = Dark Green
11 = Light Blue
12 = Blue
13 = Dark Blue
14 = Indigo
15 = Purple
16 = Black
17 = Brown
18 = Grey

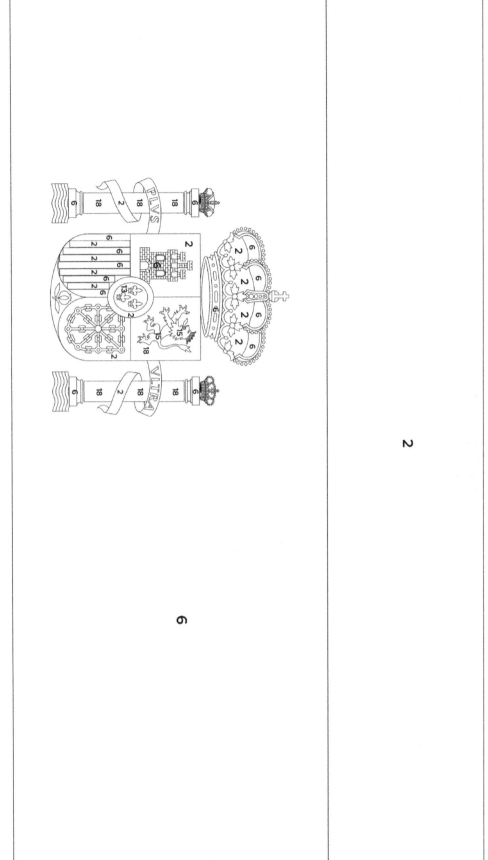

SWEDEN

12		12
	6	
12		12

1 = Maroon 3 = Red 5 = Light Orange 7 = Light Yellow 9 = Green 11 = Light Blue 13 = Dark Blue 15 = Purple 17 = Brown
2 = Dark Red 4 = Orange 6 = Golden Yellow 8 = Light Green 10 = Dark Green 12 = Blue 14 = Indigo 16 = Black 18 = Grey

SWITZERLAND

UKRAINE

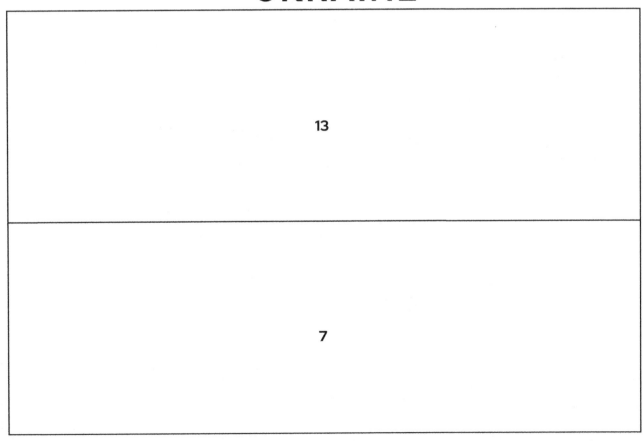

1 = Maroon	4 = Orange	7 = Light Yellow	10 = Dark Green	13 = Dark Blue	16 = Black
2 = Dark Red	5 = Light Orange	8 = Light Green	11 = Light Blue	14 = Indigo	17 = Grey
3 = Red	6 = Golden Yellow	9 = Green	12 = Blue	15 = Purple	18 = Brown

UNITED KINGDOM

14

2

14

2

14

2

14

2

14

14

14

14

2

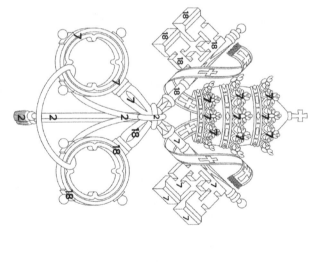

1 = Maroon   **4** = Orange   **7** = Light Yellow   **10** = Dark Green   **13** = Dark Blue   **16** = Black
2 = Dark Red   **5** = Light Orange   **8** = Light Green   **11** = Light Blue   **14** = Indigo   **17** = Grey
3 = Red   **6** = Golden Yellow   **9** = Green   **12** = Blue   **15** = Purple   **18** = Brown

AFRICAN COUNTRY FLAGS

ALGERIA

1 = Maroon
2 = Dark Red
3 = Red
4 = Orange
5 = Light Orange
6 = Golden Yellow
7 = Light Yellow
8 = Light Green
9 = Green
10 = Dark Green
11 = Light Blue
12 = Blue
13 = Dark Blue
14 = Indigo
15 = Purple
16 = Black
17 = Brown
18 = Grey

10

2

2

ANGOLA

2

6

6

6

6

6

16

BENIN

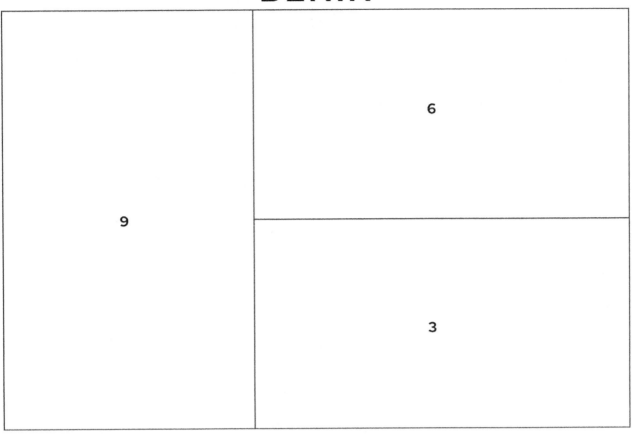

	6
9	
	3

BOTSWANA

| 11 |
| 16 |
| 11 |

1 = Maroon   **4** = Orange   **7 =** Light Yellow   **10** = Dark Green   **13** = Dark Blue   **16** = Black
2 = Dark Red   **5** = Light Orange   **8 =** Light Green   **11** = Light Blue   **14** = Indigo   **17** = Grey
3 = Red   **6** = Golden Yellow   **9** = Green   **12** = Blue   **15** = Purple   **18** = Brown

BURKINA FASO

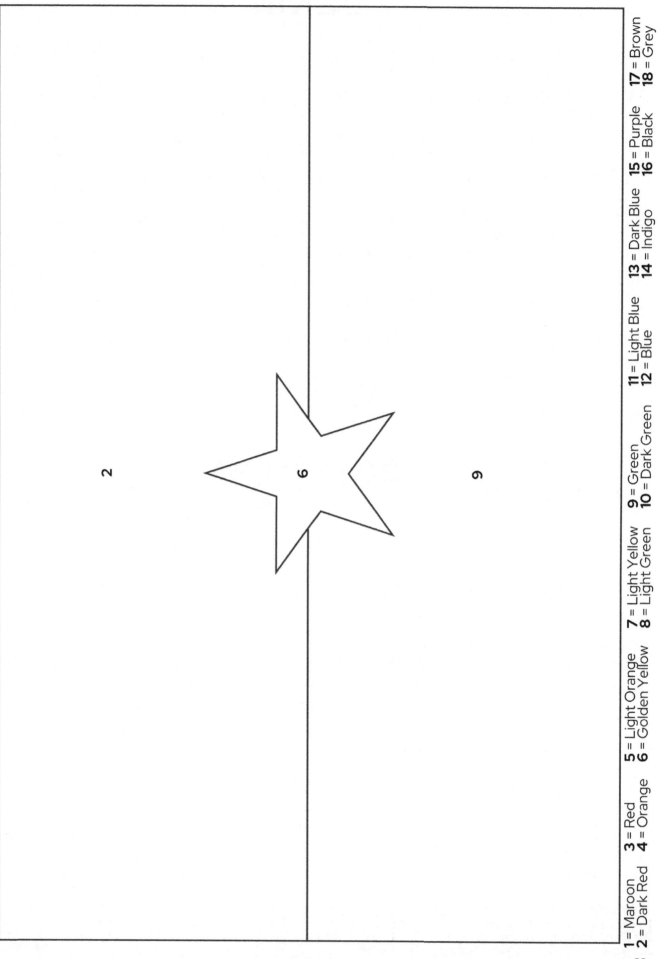

2

6

9

BURUNDI

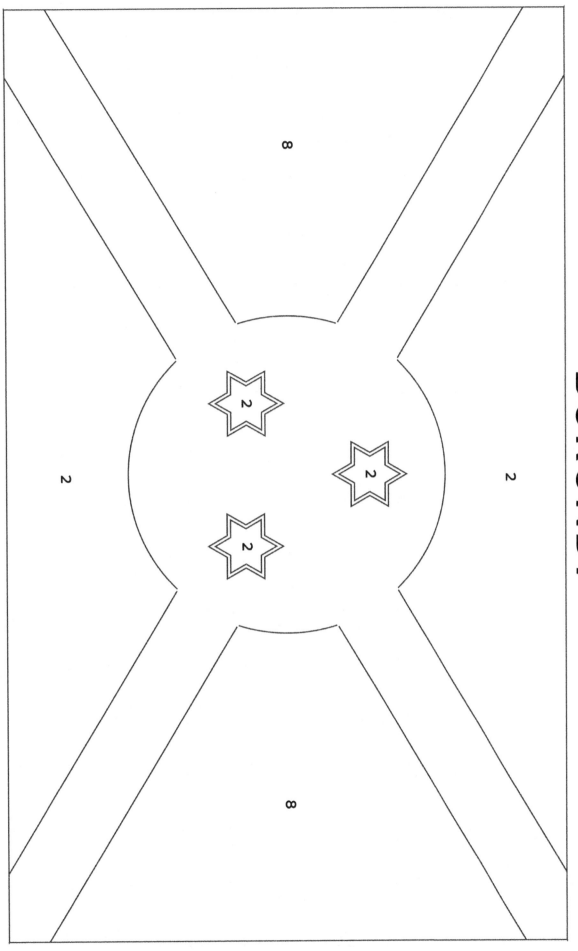

1 = Maroon
2 = Dark Red

3 = Red
4 = Orange

5 = Light Orange
6 = Golden Yellow

7 = Light Yellow
8 = Light Green

9 = Green
10 = Dark Green

11 = Light Blue
12 = Blue

13 = Dark Blue
14 = Indigo

15 = Purple
16 = Black

17 = Brown
18 = Grey

CAMEROON

2

10

6

6

CAPE VERDE

85

CENTRAL AFRICAN REPUBLIC

14

14

2

6

9

9

6

6

1 = Maroon
2 = Dark Red
3 = Red
4 = Orange
5 = Light Orange
6 = Golden Yellow
7 = Light Yellow
8 = Light Green
9 = Green
10 = Dark Green
11 = Light Blue
12 = Blue
13 = Dark Blue
14 = Indigo
15 = Purple
16 = Black
17 = Brown
18 = Grey

1 = Maroon
2 = Dark Red
3 = Red
4 = Orange
5 = Light Orange
6 = Golden Yellow
7 = Light Yellow
8 = Light Green
9 = Green
10 = Dark Green
11 = Light Blue
12 = Blue
13 = Dark Blue
14 = Indigo
15 = Purple
16 = Black
17 = Brown
18 = Grey

14

6

2

COMOROS

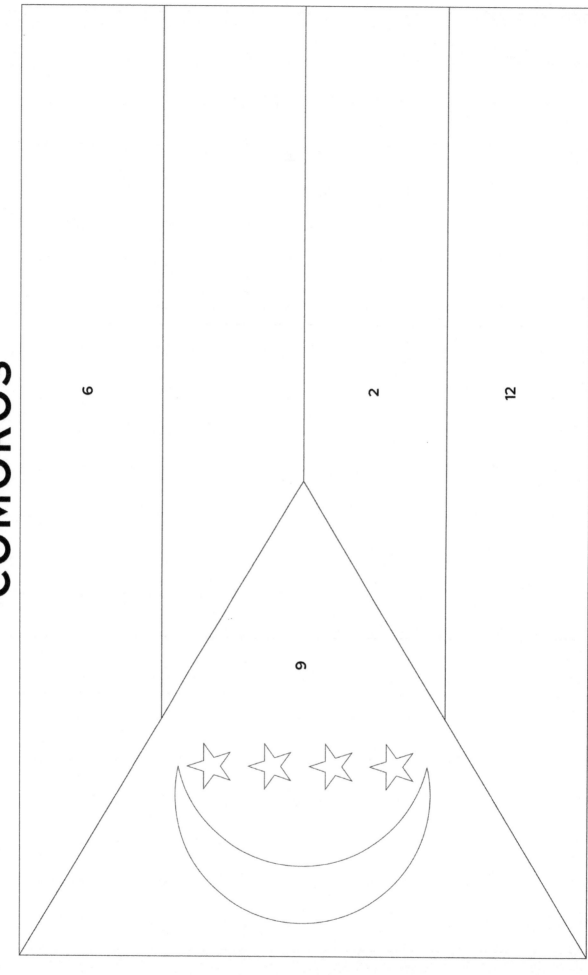

6

2

12

9

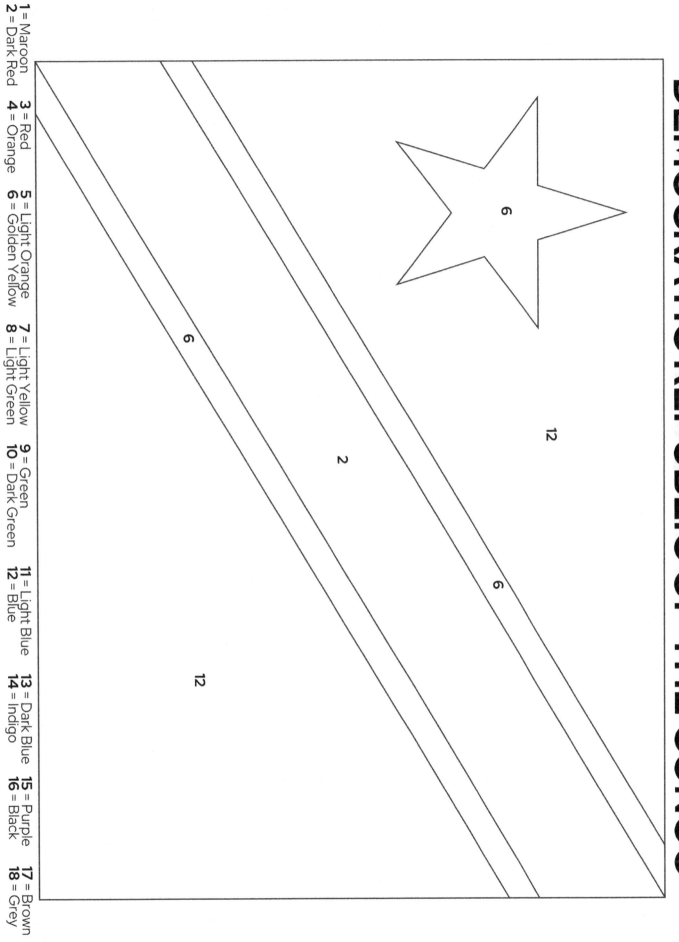

1 = Maroon
2 = Dark Red
3 = Red
4 = Orange
5 = Light Orange
6 = Golden Yellow
7 = Light Yellow
8 = Light Green
9 = Green
10 = Dark Green
11 = Light Blue
12 = Blue
13 = Dark Blue
14 = Indigo
15 = Purple
16 = Black
17 = Brown
18 = Grey

DJIBOUTI

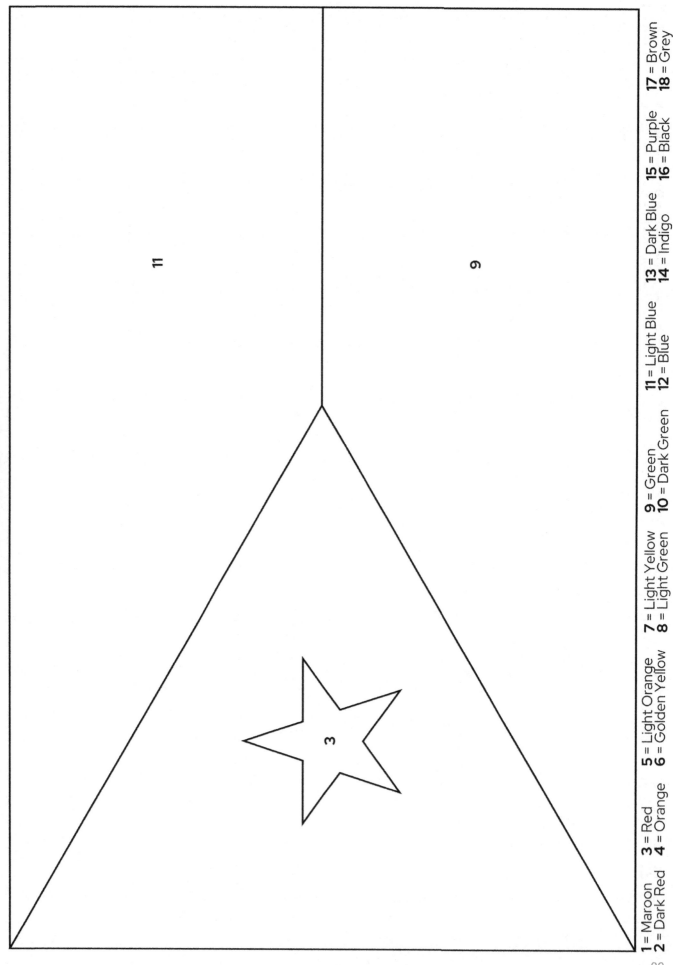

1 = Maroon
2 = Dark Red

3 = Red
4 = Orange

5 = Light Orange
6 = Golden Yellow

7 = Light Yellow
8 = Light Green

9 = Green
10 = Dark Green

11 = Light Blue
12 = Blue

13 = Dark Blue
14 = Indigo

15 = Purple
16 = Black

17 = Brown
18 = Grey

EGYPT

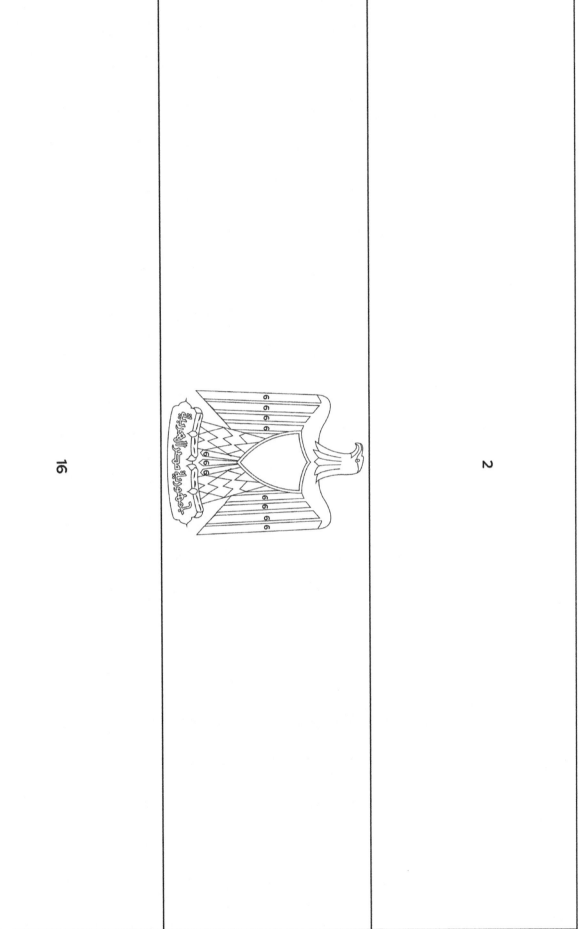

2

16

1 = Maroon   3 = Red   5 = Light Orange   7 = Light Yellow   9 = Green   11 = Light Blue   13 = Dark Blue   15 = Purple   17 = Brown
2 = Dark Red   4 = Orange   6 = Golden Yellow   8 = Light Green   10 = Dark Green   12 = Blue   14 = Indigo   16 = Black   18 = Grey

EQUATORIAL GUINEA

9

3

12

6 (on coat of arms)

7 (stars)

17 (tree trunk)

UNIDAD · PAZ · JUSTICIA

1 = Maroon
2 = Dark Red
3 = Red
4 = Orange
5 = Light Orange
6 = Golden Yellow
7 = Light Yellow
8 = Light Green
9 = Green
10 = Dark Green
11 = Light Blue
12 = Blue
13 = Dark Blue
14 = Indigo
15 = Purple
16 = Black
17 = Brown
18 = Grey

ERITREA

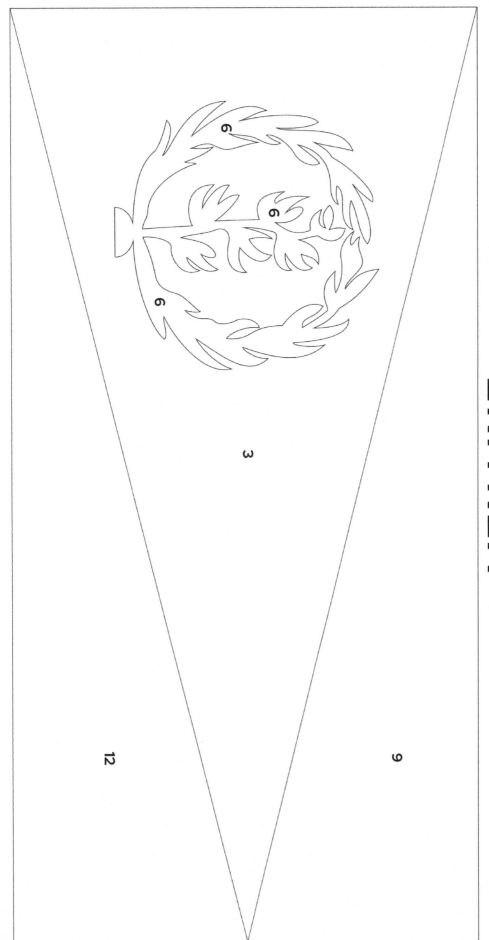

1 = Maroon
2 = Dark Red
3 = Red
4 = Orange
5 = Light Orange
6 = Golden Yellow
7 = Light Yellow
8 = Light Green
9 = Green
10 = Dark Green
11 = Light Blue
12 = Blue
13 = Dark Blue
14 = Indigo
15 = Purple
16 = Black
17 = Brown
18 = Grey

93

ESWATINI

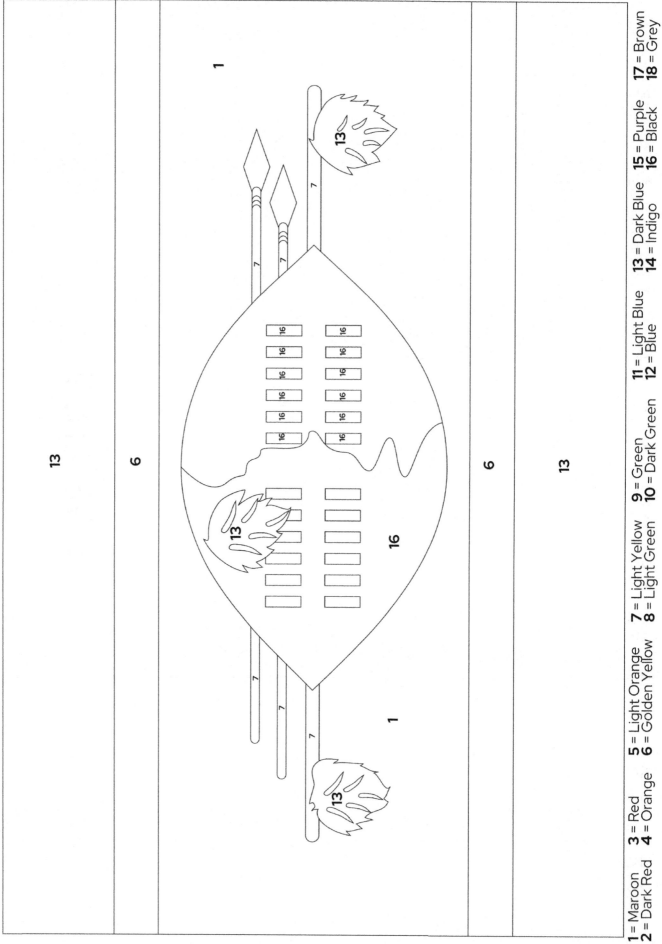

94

ETHIOPIA

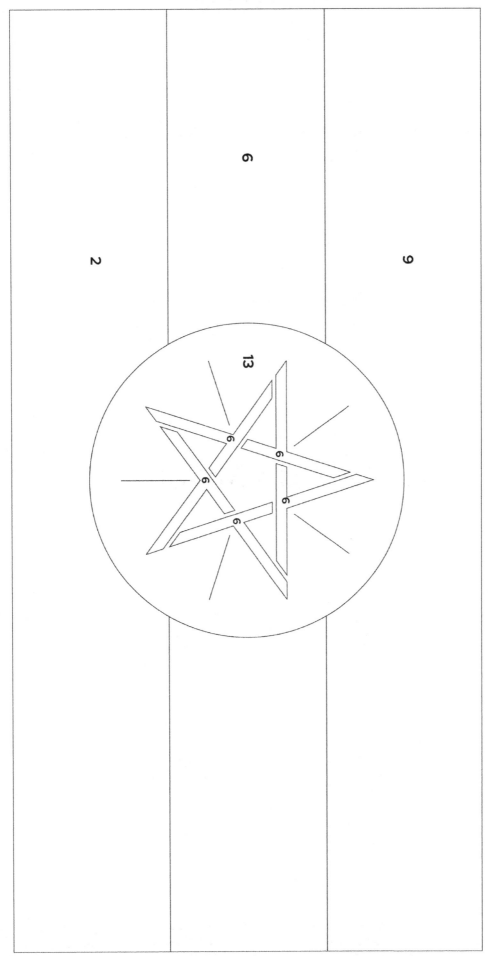

GABON

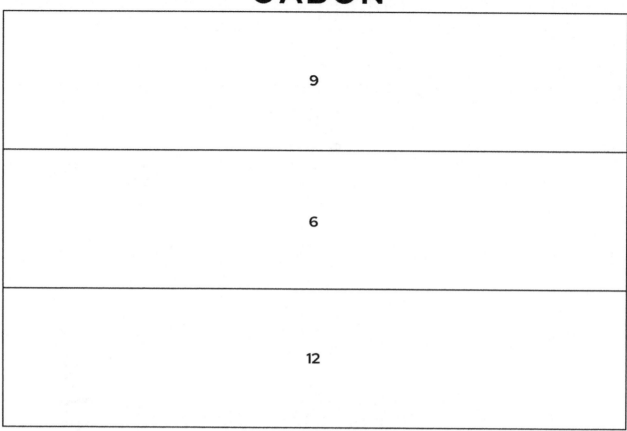

9

6

12

THE GAMBIA

2

14

10

1 = Maroon	**4** = Orange	**7** = Light Yellow	**10** = Dark Green	**13** = Dark Blue	**16** = Black
2 = Dark Red	**5** = Light Orange	**8** = Light Green	**11** = Light Blue	**14** = Indigo	**17** = Grey
3 = Red	**6** = Golden Yellow	**9** = Green	**12** = Blue	**15** = Purple	**18** = Brown

GHANA

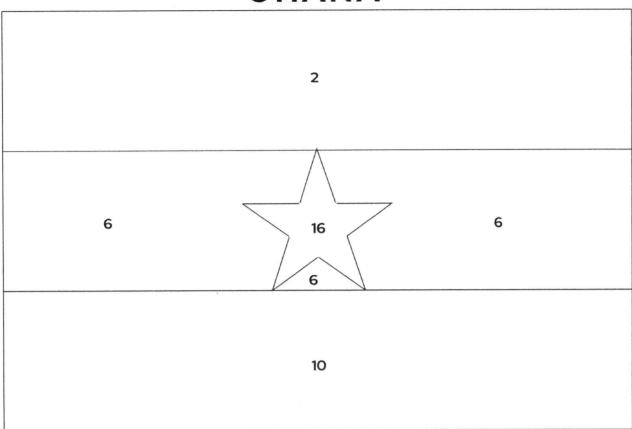

GUINEA

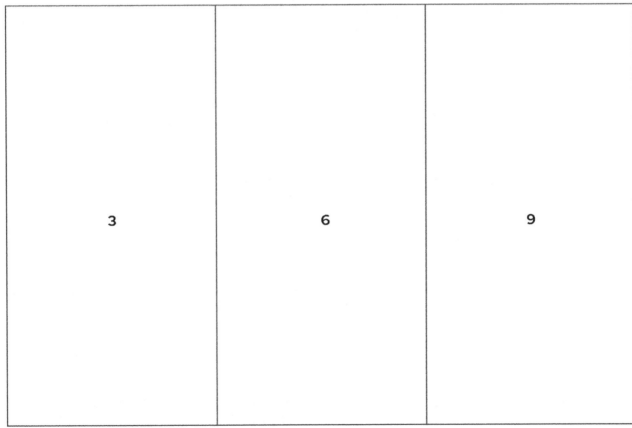

1 = Maroon   4 = Orange   7 = Light Yellow   10 = Dark Green   13 = Dark Blue   16 = Black
2 = Dark Red   5 = Light Orange   8 = Light Green   11 = Light Blue   14 = Indigo   17 = Grey
3 = Red   6 = Golden Yellow   9 = Green   12 = Blue   15 = Purple   18 = Brown

GUINEA-BISSAU

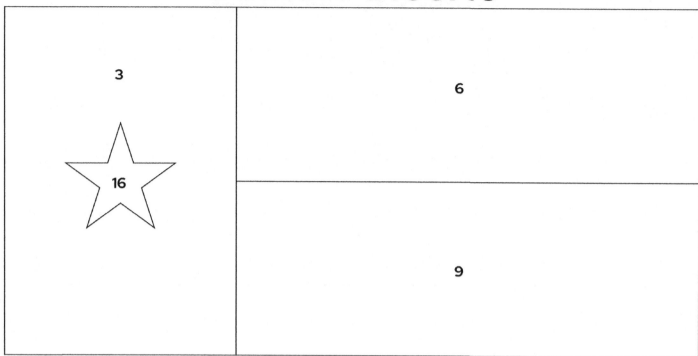

IVORY COAST

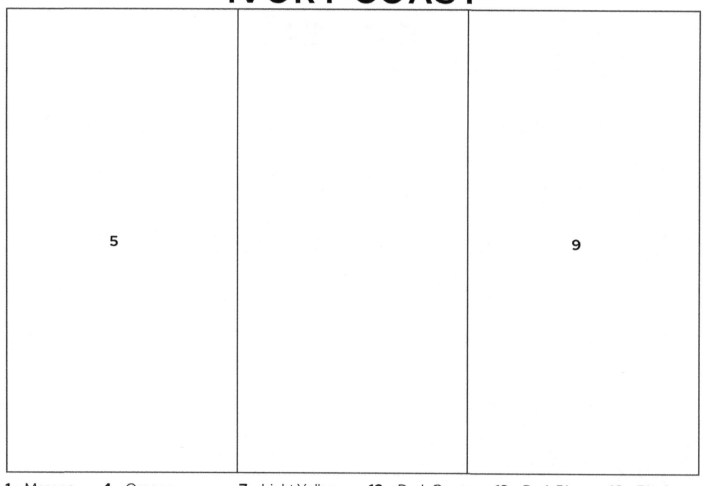

1 = Maroon   4 = Orange   7 = Light Yellow   10 = Dark Green   13 = Dark Blue   16 = Black
2 = Dark Red   5 = Light Orange   8 = Light Green   11 = Light Blue   14 = Indigo   17 = Grey
3 = Red   6 = Golden Yellow 9 = Green   12 = Blue   15 = Purple   18 = Brown

KENYA

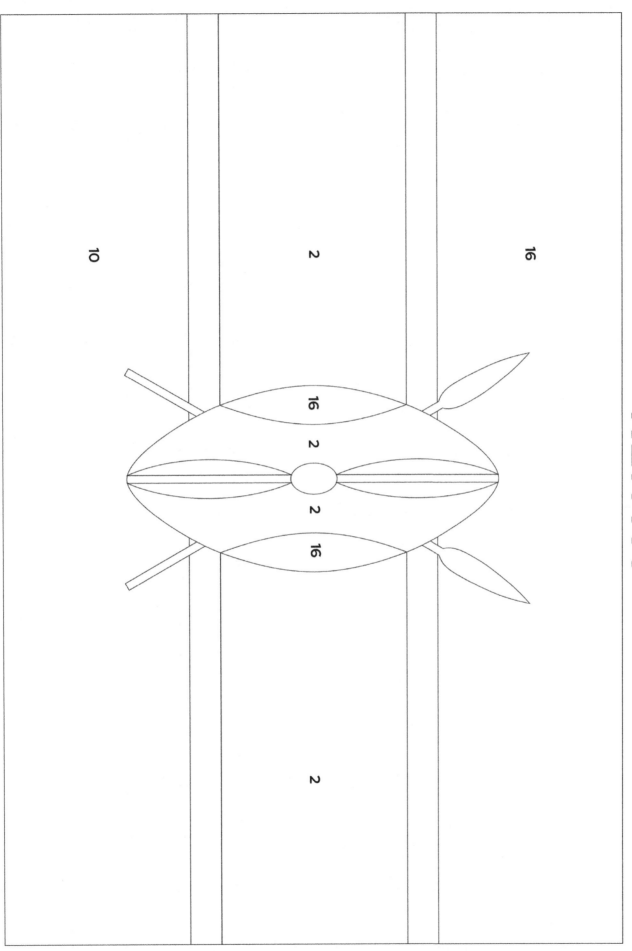

1 = Maroon   **3** = Red   **5** = Light Orange   **7** = Light Yellow   **9** = Green   **11** = Light Blue   **13** = Dark Blue   **15** = Purple   **17** = Brown
2 = Dark Red   **4** = Orange   **6** = Golden Yellow   **8** = Light Green   **10** = Dark Green   **12** = Blue   **14** = Indigo   **16** = Black   **18** = Grey

LESOTHO

13

16

9

LIBERIA

1 = Maroon
2 = Dark Red
3 = Red
4 = Orange
5 = Light Orange
6 = Golden Yellow
7 = Light Yellow
8 = Light Green
9 = Green
10 = Dark Green
11 = Light Blue
12 = Blue
13 = Dark Blue
14 = Indigo
15 = Purple
16 = Black
17 = Brown
18 = Grey

14

2

2

2

2

2

2

2

2

LIBYA

3

16

9

1 = Maroon
2 = Dark Red
3 = Red
4 = Orange
5 = Light Orange
6 = Golden Yellow
7 = Light Yellow
8 = Light Green
9 = Green
10 = Dark Green
11 = Light Blue
12 = Blue
13 = Dark Blue
14 = Indigo
15 = Purple
16 = Black
17 = Brown
18 = Grey

9

3

MALAWI

16

2

16

2

9

MALI

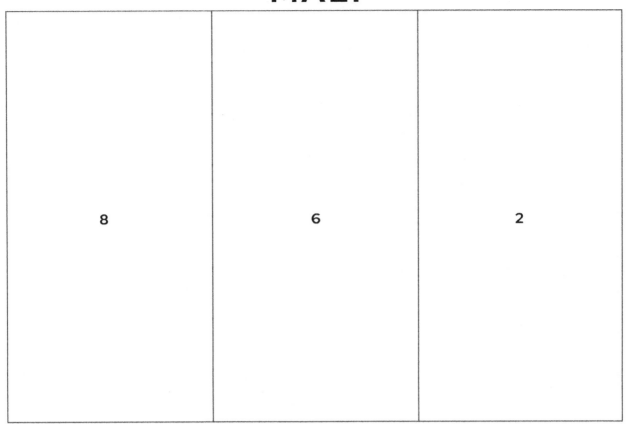

MAURITANIA

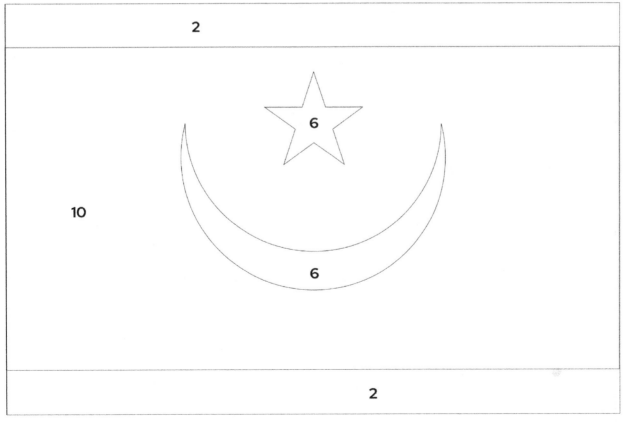

1 = Maroon	**4** = Orange	**7** = Light Yellow	**10** = Dark Green	**13** = Dark Blue	**16** = Black
2 = Dark Red	**5** = Light Orange	**8** = Light Green	**11** = Light Blue	**14** = Indigo	**17** = Grey
3 = Red	**6** = Golden Yellow	**9** = Green	**12** = Blue	**15** = Purple	**18** = Brown

MAURITIUS

3	14	6	9

1 = Maroon
2 = Dark Red
3 = Red
4 = Orange
5 = Light Orange
6 = Golden Yellow
7 = Light Yellow
8 = Light Green
9 = Green
10 = Dark Green
11 = Light Blue
12 = Blue
13 = Dark Blue
14 = Indigo
15 = Purple
16 = Black
17 = Brown
18 = Grey

2

MOZAMBIQUE

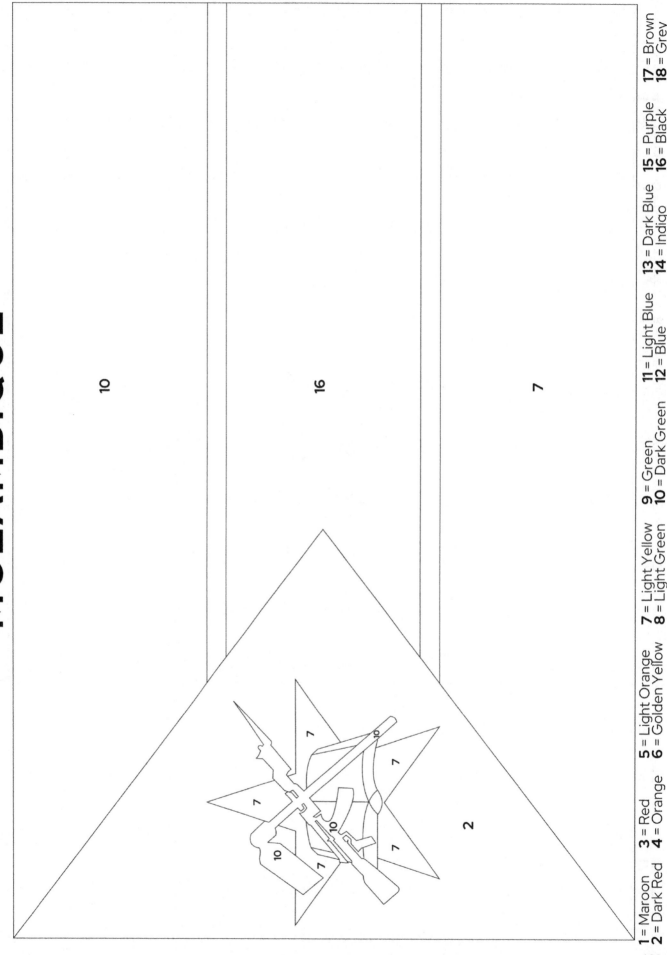

NAMIBIA

1 = Maroon
2 = Dark Red

3 = Red
4 = Orange

5 = Light Orange
6 = Golden Yellow

7 = Light Yellow
8 = Light Green

9 = Green
10 = Dark Green

11 = Light Blue
12 = Blue

13 = Dark Blue
14 = Indigo

15 = Purple
16 = Black

17 = Brown
18 = Grey

14

6

2

9

NIGER

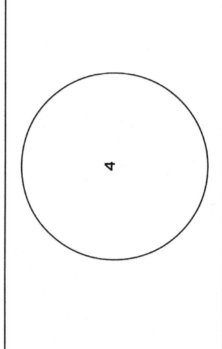

4

4

8

1 = Maroon   4 = Orange   7 = Light Yellow   10 = Dark Green   13 = Dark Blue   16 = Black
2 = Dark Red   5 = Light Orange   8 = Light Green   11 = Light Blue   14 = Indigo   17 = Grey
3 = Red   6 = Golden Yellow   9 = Green   12 = Blue   15 = Purple   18 = Brown

110

NIGERIA

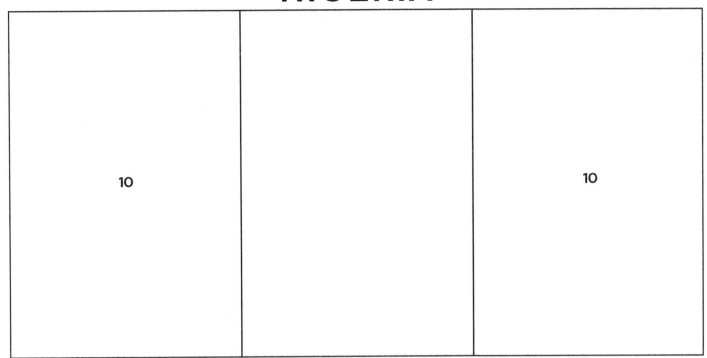

REPUBLIC OF THE CONGO

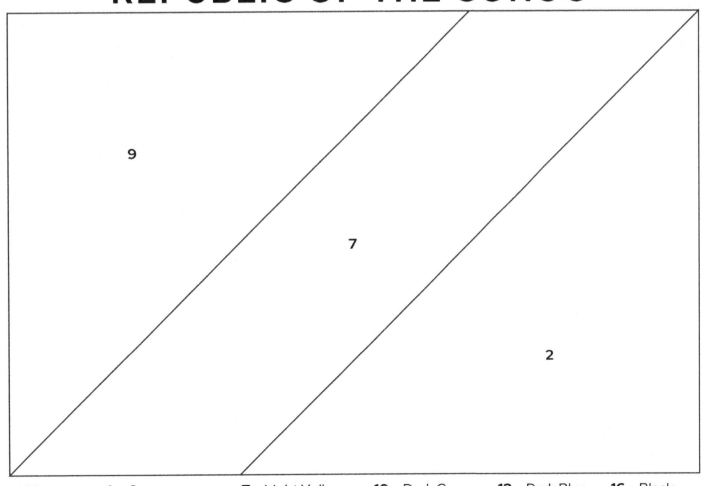

1 = Maroon   **4** = Orange   **7** = Light Yellow   **10** = Dark Green   **13** = Dark Blue   **16** = Black
2 = Dark Red   **5** = Light Orange   **8** = Light Green   **11** = Light Blue   **14** = Indigo   **17** = Grey
3 = Red   **6** = Golden Yellow   **9** = Green   **12** = Blue   **15** = Purple   **18** = Brown

RWANDA

11

7

10

1 = Maroon   3 = Red   5 = Light Orange   7 = Light Yellow   9 = Green   11 = Light Blue   13 = Dark Blue   15 = Purple   17 = Brown
2 = Dark Red   4 = Orange   6 = Golden Yellow   8 = Light Green   10 = Dark Green   12 = Blue   14 = Indigo   16 = Black   18 = Grey

SAO TOME AND PRINCIPE

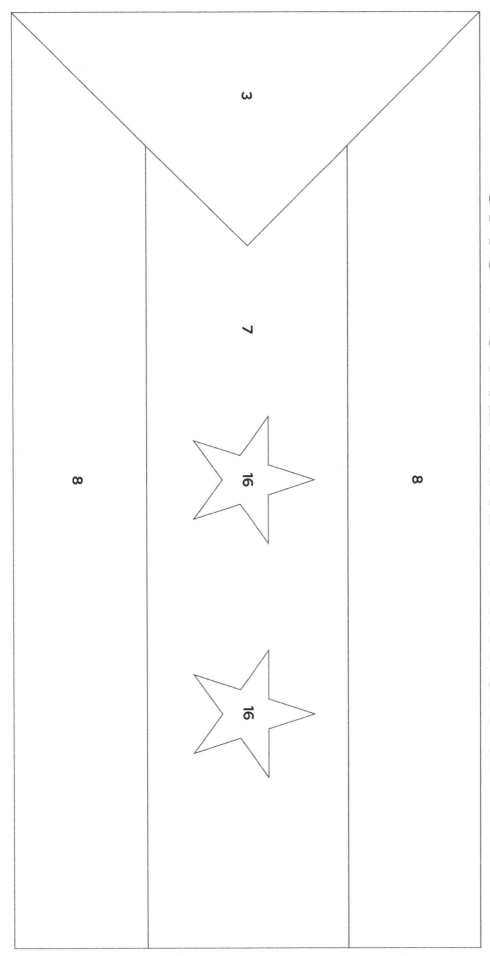

1 = Maroon
2 = Dark Red

3 = Red
4 = Orange

5 = Light Orange
6 = Golden Yellow

7 = Light Yellow
8 = Light Green

9 = Green
10 = Dark Green

11 = Light Blue
12 = Blue

13 = Dark Blue
14 = Indigo

15 = Purple
16 = Black

17 = Brown
18 = Grey

SENEGAL

7

3

6

9

SEYCHELLES

14

6

3

10

SIERRA LEONE

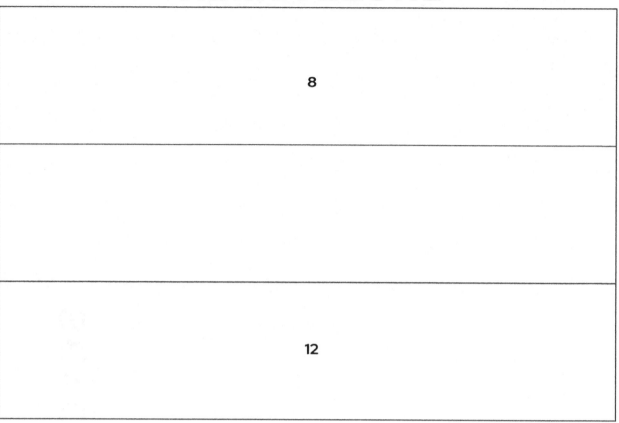

SOMALIA

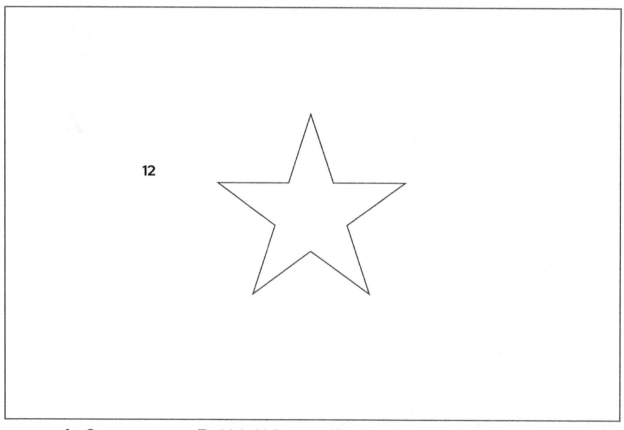

1 = Maroon   4 = Orange   7 = Light Yellow   10 = Dark Green   13 = Dark Blue   16 = Black
2 = Dark Red   5 = Light Orange   8 = Light Green   11 = Light Blue   14 = Indigo   17 = Grey
3 = Red   6 = Golden Yellow   9 = Green   12 = Blue   15 = Purple   18 = Brown

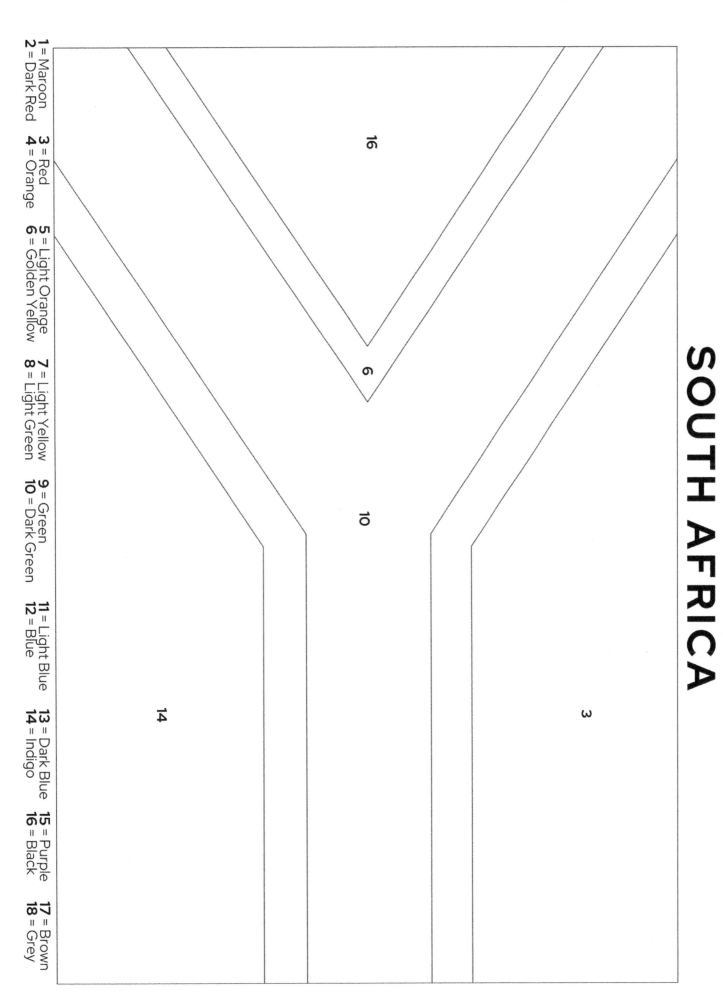

SOUTH AFRICA

1 = Maroon
2 = Dark Red

3 = Red
4 = Orange

5 = Light Orange
6 = Golden Yellow

7 = Light Yellow
8 = Light Green

9 = Green
10 = Dark Green

11 = Light Blue
12 = Blue

13 = Dark Blue
14 = Indigo

15 = Purple
16 = Black

17 = Brown
18 = Grey

SOUTH SUDAN

16

3

9

13

7

1 = Maroon
2 = Dark Red

3 = Red
4 = Orange

5 = Light Orange
6 = Golden Yellow

7 = Light Yellow
8 = Light Green

9 = Green
10 = Dark Green

11 = Light Blue
12 = Blue

13 = Dark Blue
14 = Indigo

15 = Purple
16 = Black

17 = Brown
18 = Grey

SUDAN

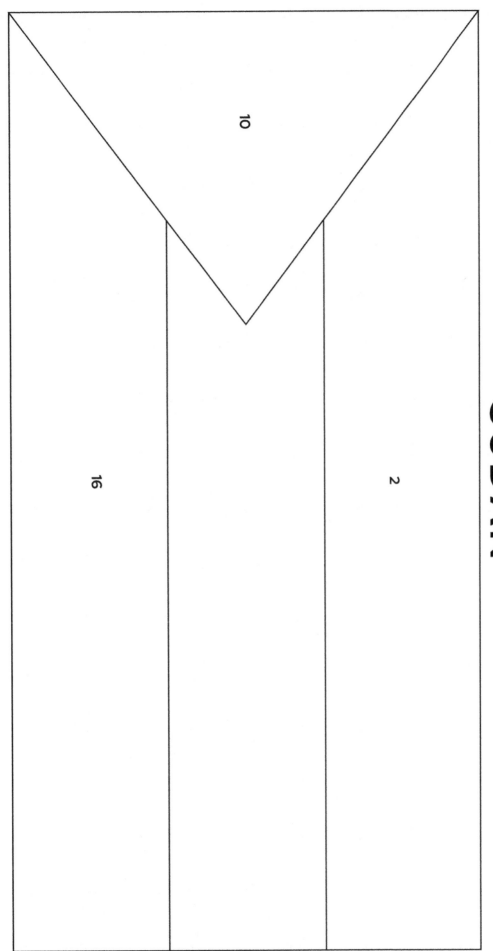

1 = Maroon
2 = Dark Red

3 = Red
4 = Orange

5 = Light Orange
6 = Golden Yellow

7 = Light Yellow
8 = Light Green

9 = Green
10 = Dark Green

11 = Light Blue
12 = Blue

13 = Dark Blue
14 = Indigo

15 = Purple
16 = Black

17 = Brown
18 = Grey

TANZANIA

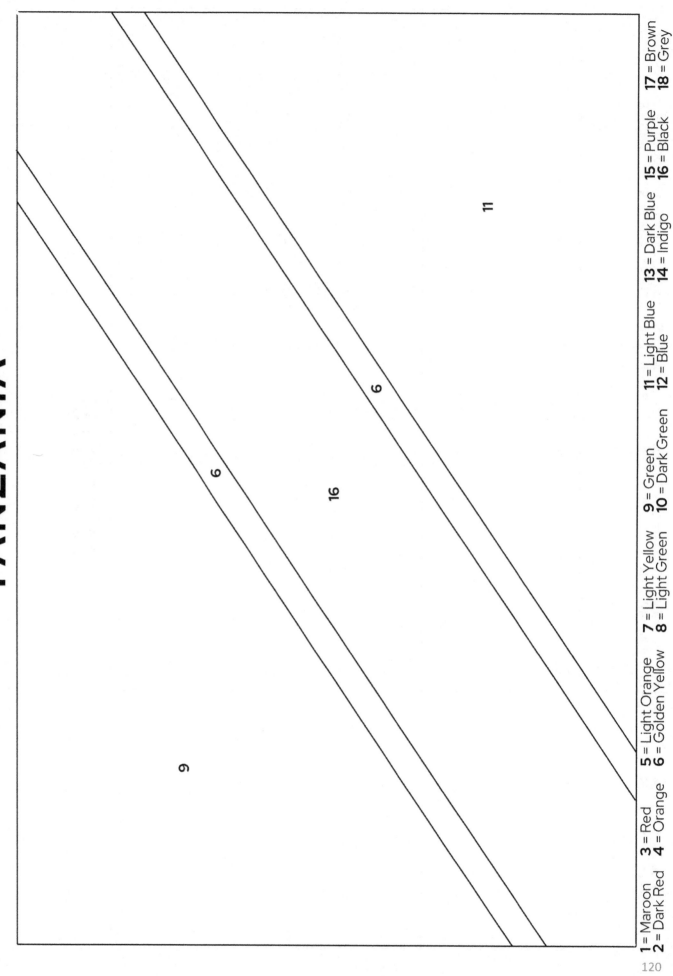

9

16

6

11

6

TOGO

1 = Maroon
2 = Dark Red
3 = Red
4 = Orange
5 = Light Orange
6 = Golden Yellow
7 = Light Yellow
8 = Light Green
9 = Green
10 = Dark Green
11 = Light Blue
12 = Blue
13 = Dark Blue
14 = Indigo
15 = Purple
16 = Black
17 = Brown
18 = Grey

2

10

6

10

6

10

TUNISIA

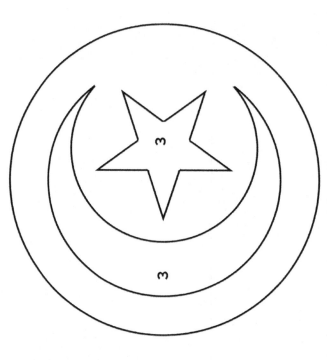

UGANDA

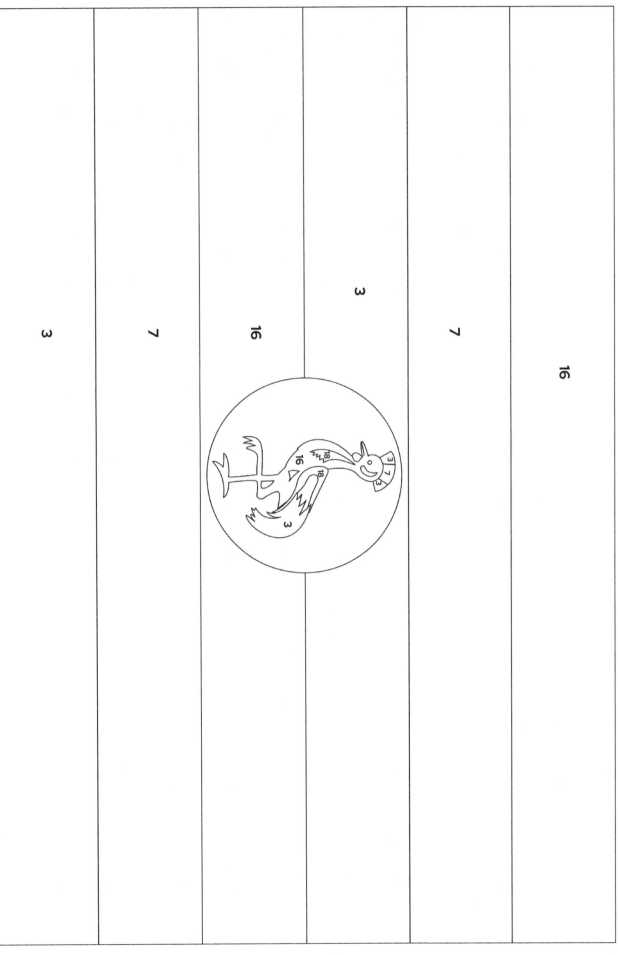

1 = Maroon
2 = Dark Red
3 = Red
4 = Orange
5 = Light Orange
6 = Golden Yellow
7 = Light Yellow
8 = Light Green
9 = Green
10 = Dark Green
11 = Light Blue
12 = Blue
13 = Dark Blue
14 = Indigo
15 = Purple
16 = Black
17 = Brown
18 = Grey

ZAMBIA

9

2   16   4

ZIMBABWE

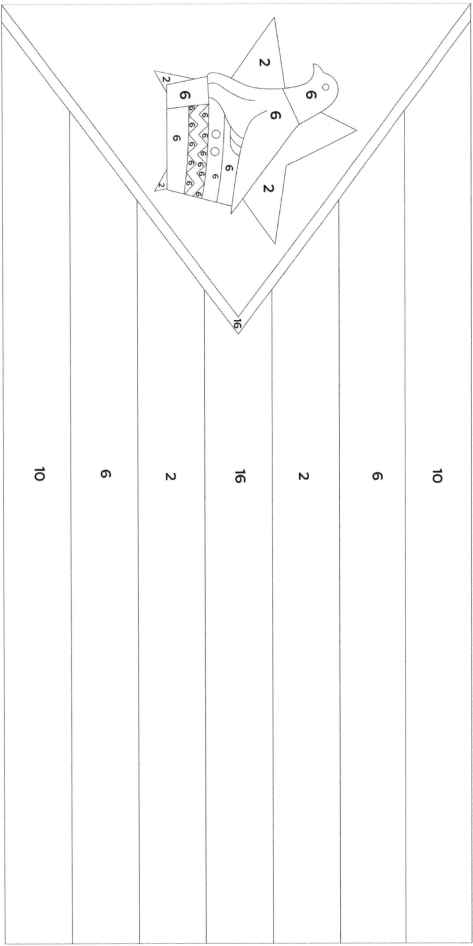

1 = Maroon
2 = Dark Red
3 = Red
4 = Orange
5 = Light Orange
6 = Golden Yellow
7 = Light Yellow
8 = Light Green
9 = Green
10 = Dark Green
11 = Light Blue
12 = Blue
13 = Dark Blue
14 = Indigo
15 = Purple
16 = Black
17 = Brown
18 = Grey

125

ASIAN
COUNTRY FLAGS

AFGHANISTAN

127

1 = Maroon
2 = Dark Red

3 = Red
4 = Orange

5 = Light Orange
6 = Golden Yellow

7 = Light Yellow
8 = Light Green

9 = Green
10 = Dark Green

11 = Light Blue
12 = Blue

13 = Dark Blue
14 = Indigo

15 = Purple
16 = Black

17 = Brown
18 = Grey

16

3

3

3

3

3

3

3

3

3

3

10

ARMENIA

3	13	6

AZERBAIJAN

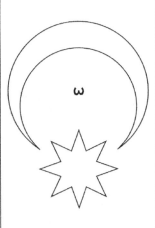

11

3

3

3

9

3

BAHRAIN

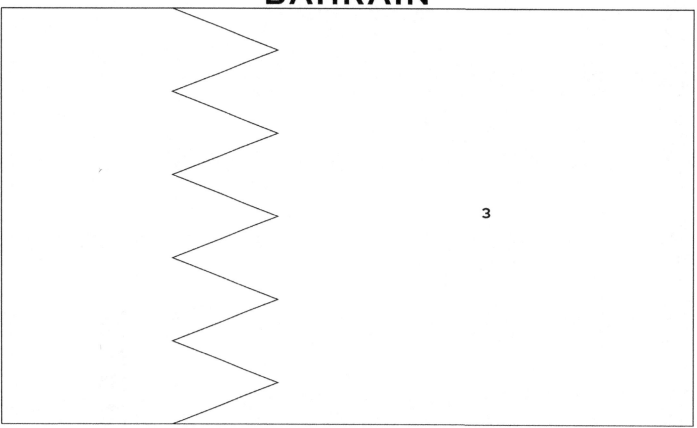

BANGLADESH

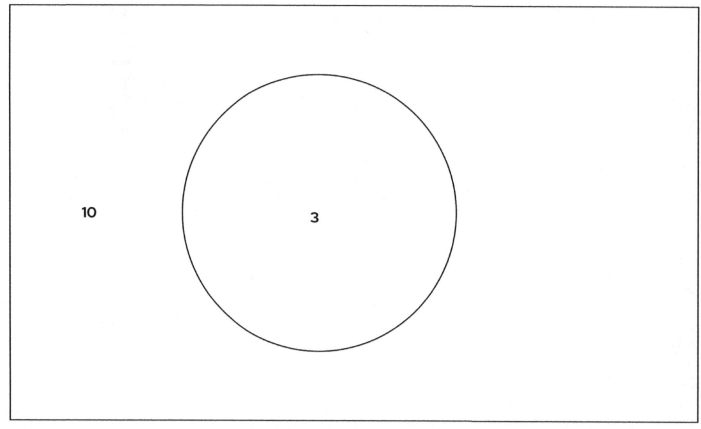

1 = Maroon   4 = Orange   7 = Light Yellow   10 = Dark Green   13 = Dark Blue   16 = Black
2 = Dark Red   5 = Light Orange   8 = Light Green   11 = Light Blue   14 = Indigo   17 = Grey
3 = Red   6 = Golden Yellow   9 = Green   12 = Blue   15 = Purple   18 = Brown

BHUTAN

1 = Maroon
2 = Dark Red
3 = Red
4 = Orange
5 = Light Orange
6 = Golden Yellow
7 = Light Yellow
8 = Light Green
9 = Green
10 = Dark Green
11 = Light Blue
12 = Blue
13 = Dark Blue
14 = Indigo
15 = Purple
16 = Black
17 = Brown
18 = Grey

BRUNEI

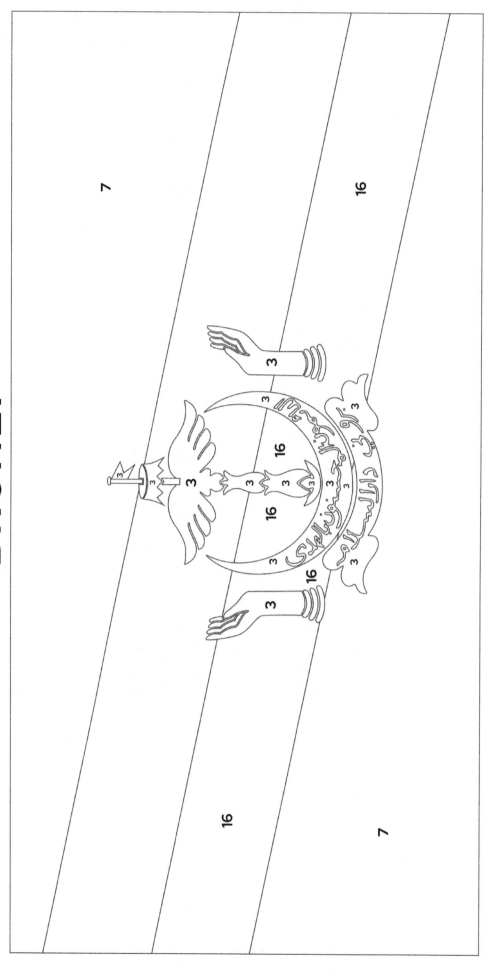

CAMBODIA

1 = Maroon
2 = Dark Red
3 = Red
4 = Orange
5 = Light Orange
6 = Golden Yellow
7 = Light Yellow
8 = Light Green
9 = Green
10 = Dark Green
11 = Light Blue
12 = Blue
13 = Dark Blue
14 = Indigo
15 = Purple
16 = Black
17 = Brown
18 = Grey

13

13

3

13

CHINA

3

6

6

6

6

6

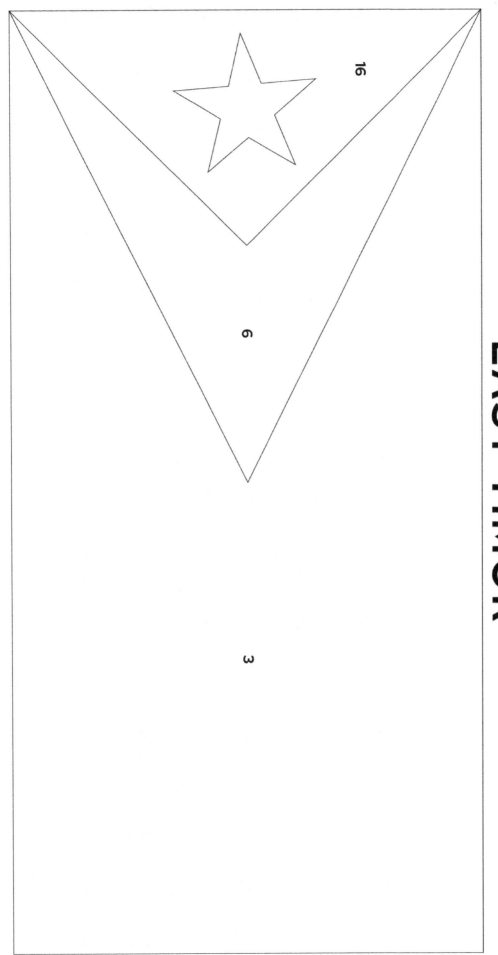

1 = Maroon
2 = Dark Red

3 = Red
4 = Orange

5 = Light Orange
6 = Golden Yellow

7 = Light Yellow
8 = Light Green

9 = Green
10 = Dark Green

11 = Light Blue
12 = Blue

13 = Dark Blue
14 = Indigo

15 = Purple
16 = Black

17 = Brown
18 = Grey

135

GEORGIA

3

3

3

3

3

1 = Maroon
2 = Dark Red

3 = Red
4 = Orange

5 = Light Orange
6 = Golden Yellow

7 = Light Yellow
8 = Light Green

9 = Green
10 = Dark Green

11 = Light Blue
12 = Blue

13 = Dark Blue
14 = Indigo

15 = Purple
16 = Black

17 = Brown
18 = Grey

INDIA

5

9

14

INDONESIA

3

IRAN

1 = Maroon
2 = Dark Red
3 = Red
4 = Orange
5 = Light Orange
6 = Golden Yellow
7 = Light Yellow
8 = Light Green
9 = Green
10 = Dark Green
11 = Light Blue
12 = Blue
13 = Dark Blue
14 = Indigo
15 = Purple
16 = Black
17 = Brown
18 = Grey

9

2

2

2

2

2

2

IRAQ

2

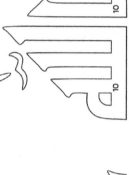

16

ISRAEL

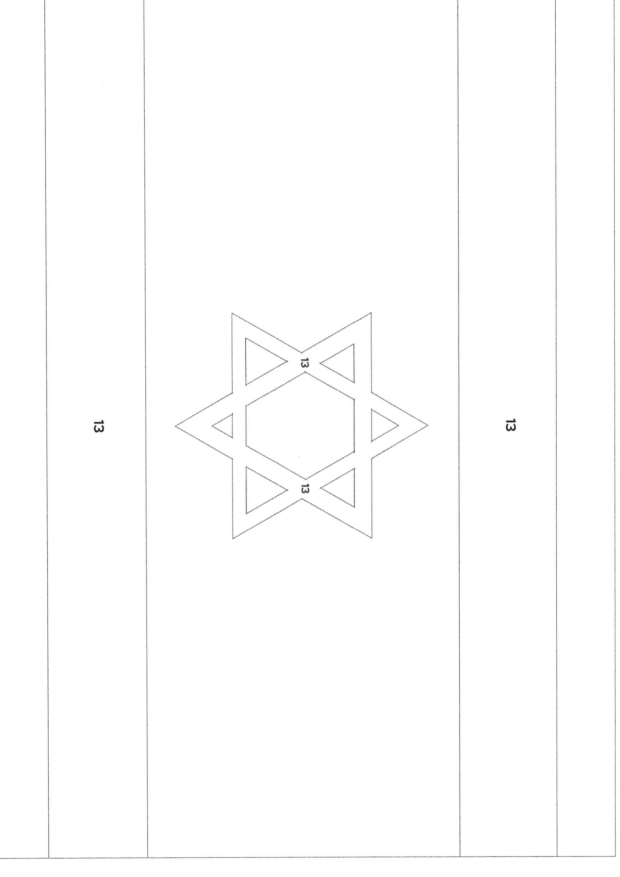

1 = Maroon
2 = Dark Red
3 = Red
4 = Orange
5 = Light Orange
6 = Golden Yellow
7 = Light Yellow
8 = Light Green
9 = Green
10 = Dark Green
11 = Light Blue
12 = Blue
13 = Dark Blue
14 = Indigo
15 = Purple
16 = Black
17 = Brown
18 = Grey

JAPAN

2

JORDAN

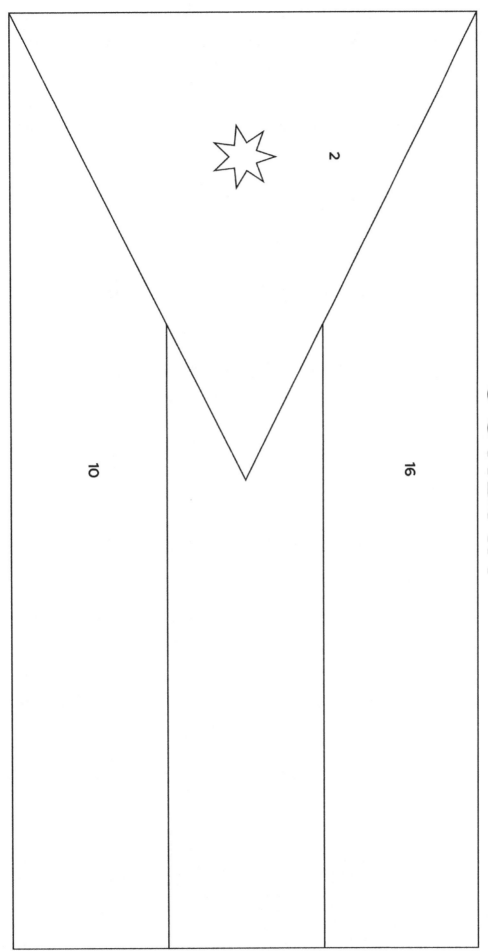

KAZAKHSTAN

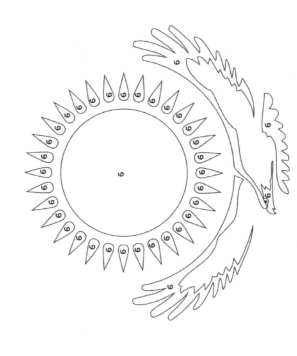

11

KUWAIT

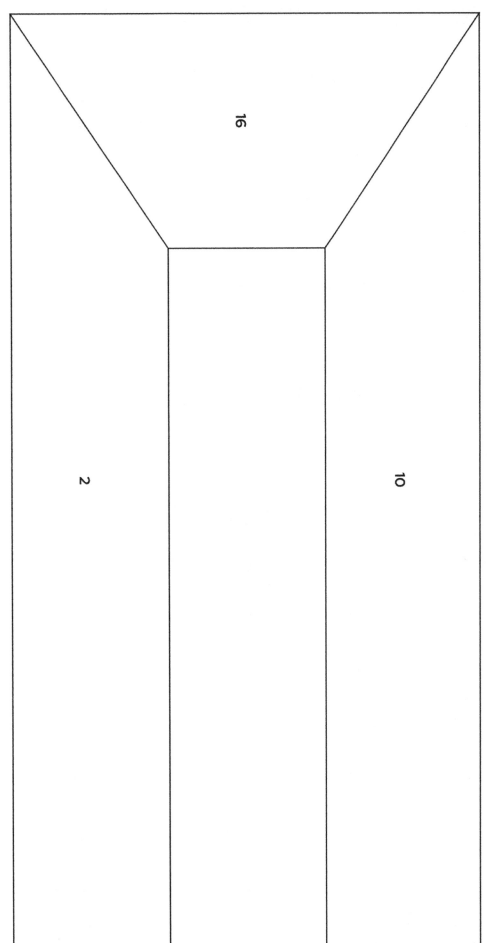

KYRGYZSTAN

3

LAOS

2

2

14

147

LEBANON

3

3

6

MALAYSIA

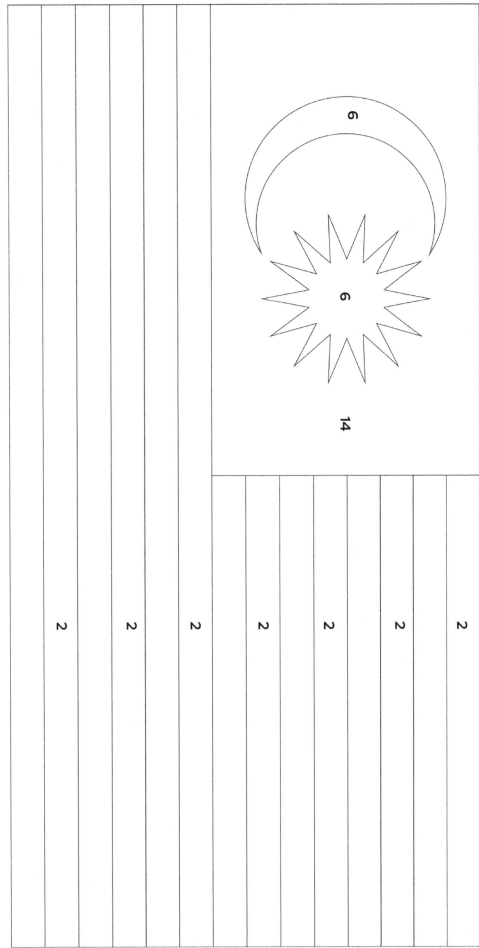

1 = Maroon
2 = Dark Red
3 = Red
4 = Orange
5 = Light Orange
6 = Golden Yellow
7 = Light Yellow
8 = Light Green
9 = Green
10 = Dark Green
11 = Light Blue
12 = Blue
13 = Dark Blue
14 = Indigo
15 = Purple
16 = Black
17 = Brown
18 = Grey

MALDIVES

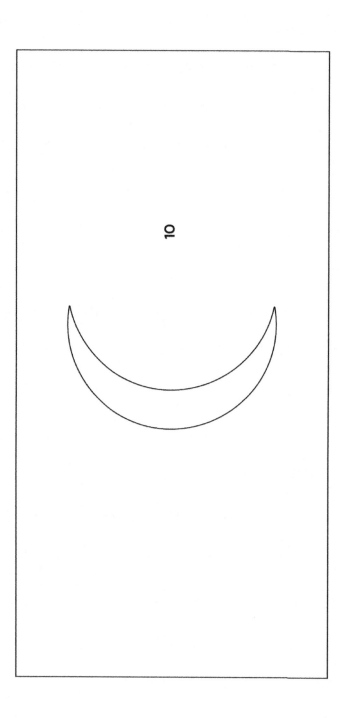

MONGOLIA

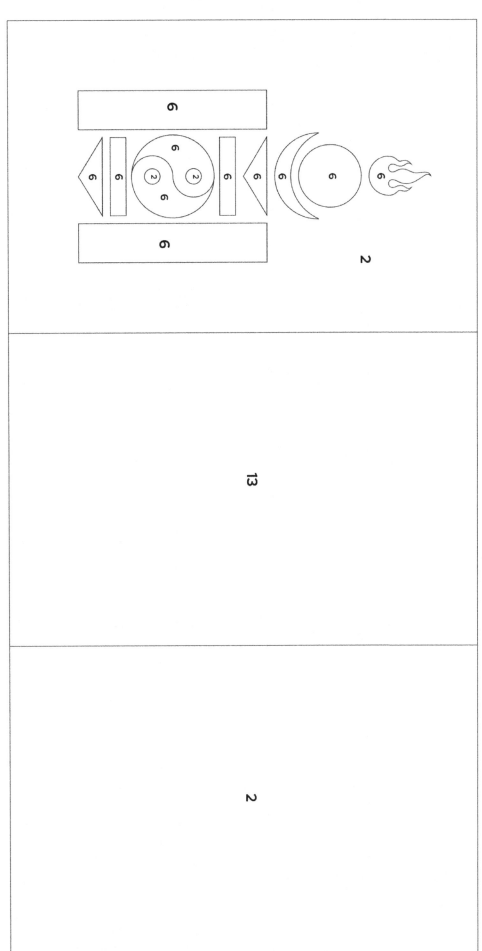

1 = Maroon
2 = Dark Red
3 = Red
4 = Orange
5 = Light Orange
6 = Golden Yellow
7 = Light Yellow
8 = Light Green
9 = Green
10 = Dark Green
11 = Light Blue
12 = Blue
13 = Dark Blue
14 = Indigo
15 = Purple
16 = Black
17 = Brown
18 = Grey

MYANMAR

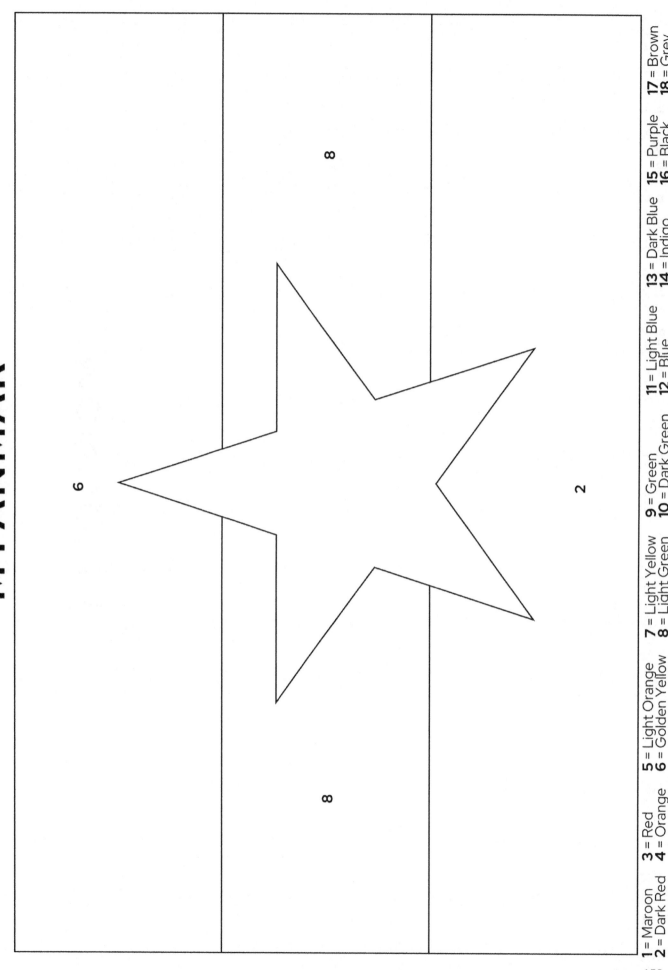

1 = Maroon
2 = Dark Red
3 = Red
4 = Orange
5 = Light Orange
6 = Golden Yellow
7 = Light Yellow
8 = Light Green
9 = Green
10 = Dark Green
11 = Light Blue
12 = Blue
13 = Dark Blue
14 = Indigo
15 = Purple
16 = Black
17 = Brown
18 = Grey

NEPAL

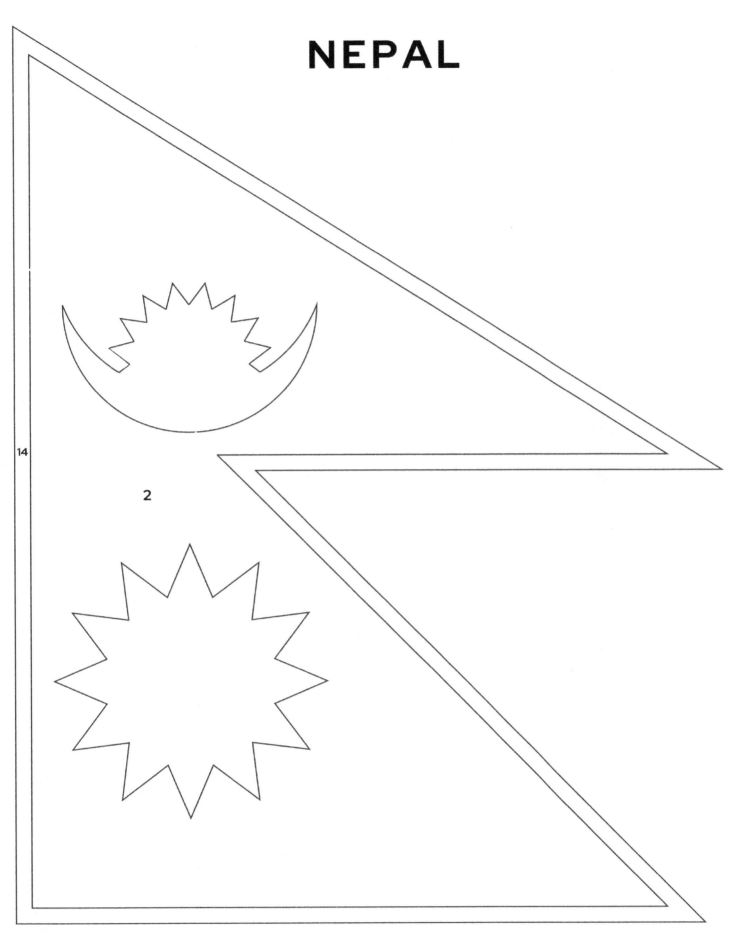

14

2

1 = Maroon   4 = Orange   7 = Light Yellow   10 = Dark Green   13 = Dark Blue   16 = Black
2 = Dark Red   5 = Light Orange   8 = Light Green   11 = Light Blue   14 = Indigo   17 = Grey
3 = Red   6 = Golden Yellow   9 = Green   12 = Blue   15 = Purple   18 = Brown

NORTH KOREA

13

3

3

3

13

OMAN

3

9

1 = Maroon
2 = Dark Red

3 = Red
4 = Orange

5 = Light Orange
6 = Golden Yellow

7 = Light Yellow
8 = Light Green

9 = Green
10 = Dark Green

11 = Light Blue
12 = Blue

13 = Dark Blue
14 = Indigo

15 = Purple
16 = Black

17 = Brown
18 = Grey

PAKISTAN

10

PALESTINE

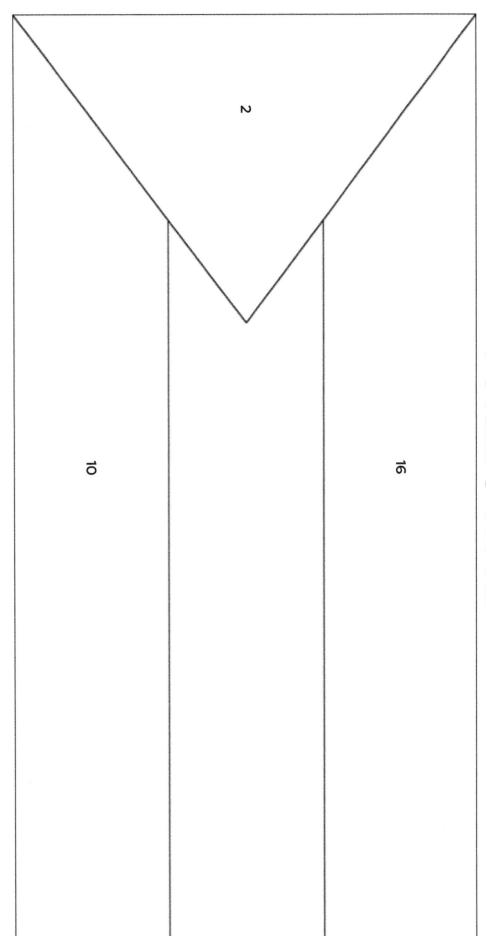

157

1 = Maroon 　3 = Red 　5 = Light Orange 　7 = Light Yellow 　9 = Green 　11 = Light Blue 　13 = Dark Blue 　15 = Purple 　17 = Brown
2 = Dark Red 　4 = Orange 　6 = Golden Yellow 　8 = Light Green 　10 = Dark Green 　12 = Blue 　14 = Indigo 　16 = Black 　18 = Grey

PHILIPPINES

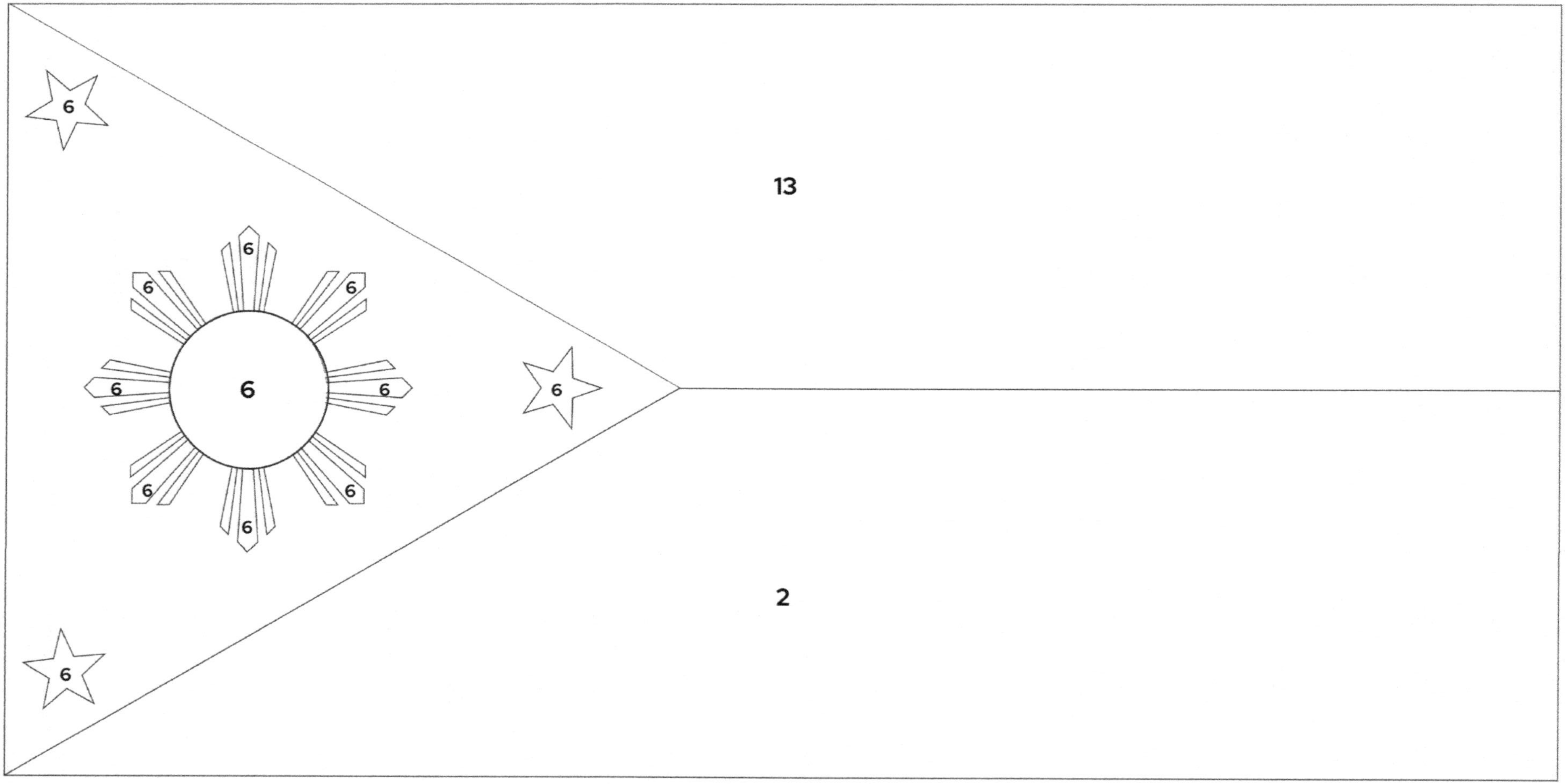

13

2

6

6

6

6

6

6

6

6

6

6

6

158

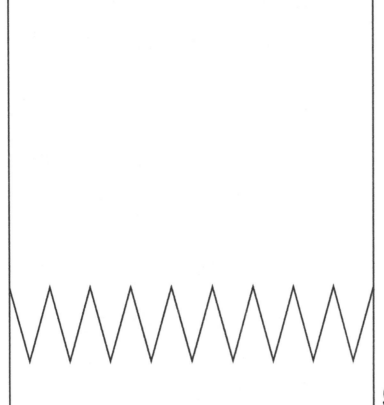

QATAR

1 = Maroon
2 = Dark Red
3 = Red
4 = Orange
5 = Light Orange
6 = Golden Yellow
7 = Light Yellow
8 = Light Green
9 = Green
10 = Dark Green
11 = Light Blue
12 = Blue
13 = Dark Blue
14 = Indigo
15 = Purple
16 = Black
17 = Brown
18 = Grey

SAUDI ARABIA

SINGAPORE

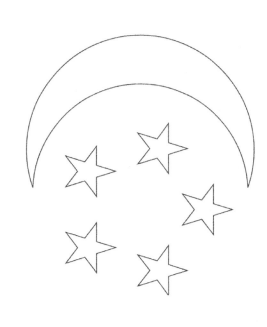

3

SOUTH KOREA

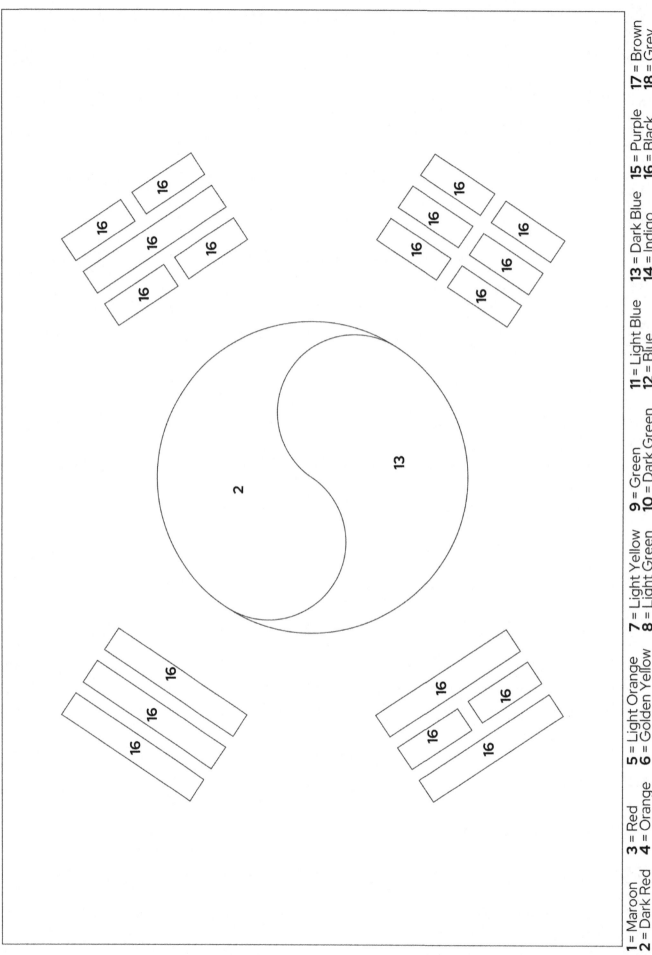

1 = Maroon   3 = Red   5 = Light Orange   7 = Light Yellow   9 = Green   11 = Light Blue   13 = Dark Blue   15 = Purple   17 = Brown
2 = Dark Red   4 = Orange   6 = Golden Yellow   8 = Light Green   10 = Dark Green   12 = Blue   14 = Indigo   16 = Black   18 = Grey

SRI LANKA

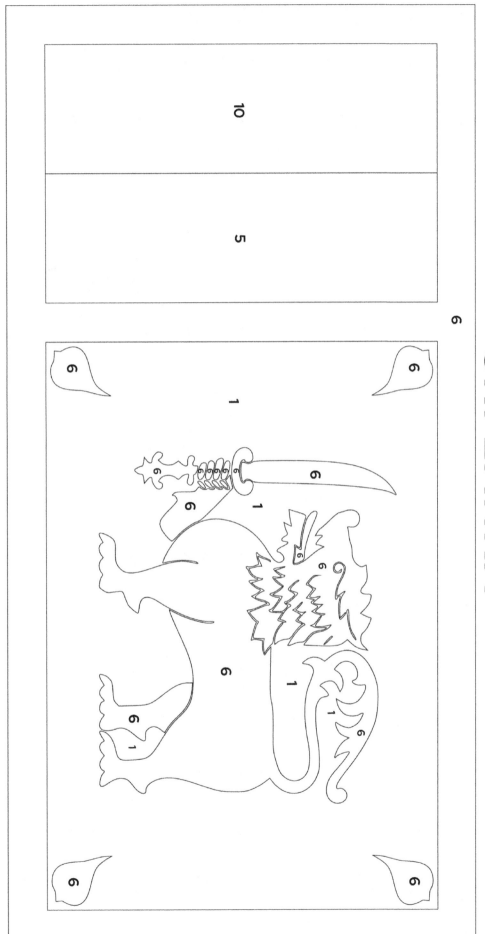

1 = Maroon
2 = Dark Red
3 = Red
4 = Orange
5 = Light Orange
6 = Golden Yellow
7 = Light Yellow
8 = Light Green
9 = Green
10 = Dark Green
11 = Light Blue
12 = Blue
13 = Dark Blue
14 = Indigo
15 = Purple
16 = Black
17 = Brown
18 = Grey

SYRIA

2

★ 10

★ 10

16

TAJIKISTAN

2

10

THAILAND

1
14
1

TURKEY

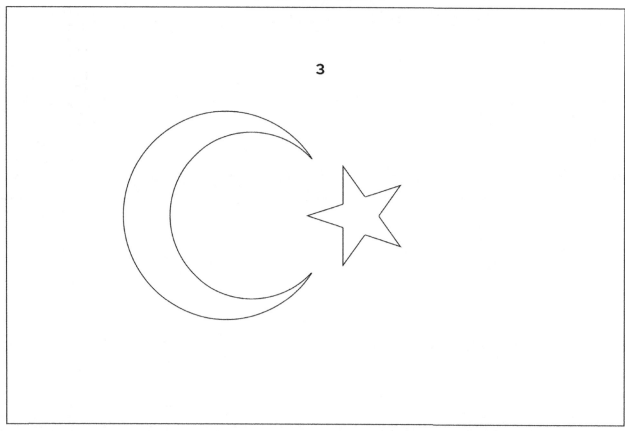

3

1 = Maroon   **4** = Orange   **7** = Light Yellow   **10** = Dark Green   **13** = Dark Blue   **16** = Black
2 = Dark Red   **5** = Light Orange   **8** = Light Green   **11** = Light Blue   **14** = Indigo   **17** = Grey
3 = Red   **6** = Golden Yellow   **9** = Green   **12** = Blue   **15** = Purple   **18** = Brown

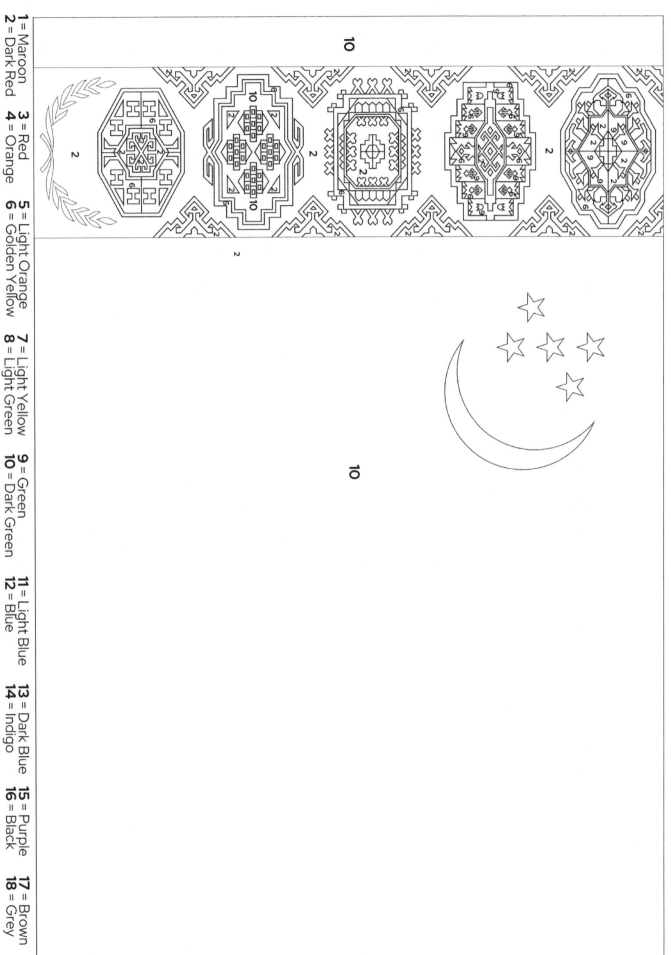

TURKMENISTAN

167

1 = Maroon
2 = Dark Red
3 = Red
4 = Orange
5 = Light Orange
6 = Golden Yellow
7 = Light Yellow
8 = Light Green
9 = Green
10 = Dark Green
11 = Light Blue
12 = Blue
13 = Dark Blue
14 = Indigo
15 = Purple
16 = Black
17 = Brown
18 = Grey

UNITED ARAB EMIRATES

10	3
16	

1 = Maroon   **3** = Red   **5** = Light Orange   **7** = Light Yellow   **9** = Green   **11** = Light Blue   **13** = Dark Blue   **15** = Purple   **17** = Brown
2 = Dark Red   **4** = Orange   **6** = Golden Yellow   **8** = Light Green   **10** = Dark Green   **12** = Blue   **14** = Indigo   **16** = Black   **18** = Grey

UZBEKISTAN

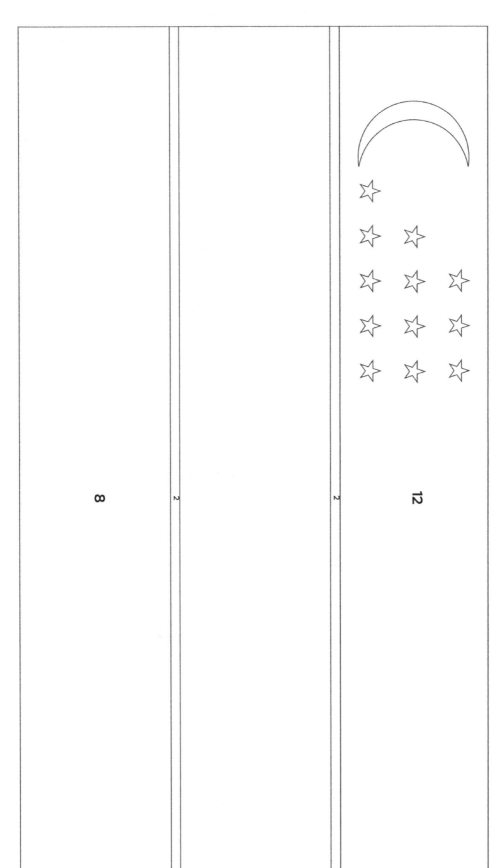

1 = Maroon
2 = Dark Red
3 = Red
4 = Orange
5 = Light Orange
6 = Golden Yellow
7 = Light Yellow
8 = Light Green
9 = Green
10 = Dark Green
11 = Light Blue
12 = Blue
13 = Dark Blue
14 = Indigo
15 = Purple
16 = Black
17 = Brown
18 = Grey

VIETNAM

3

7

YEMEN

2

16

1 = Maroon   **4** = Orange   **7 =** Light Yellow   **10** = Dark Green   **13** = Dark Blue   **16** = Black
2 = Dark Red   **5** = Light Orange   **8 =** Light Green   **11** = Light Blue   **14** = Indigo   **17** = Grey
3 = Red   **6** = Golden Yellow   **9** = Green   **12** = Blue   **15** = Purple   **18** = Brown

AUSTRALIA & OCEANIAN COUNTRY FLAGS

AUSTRALIA

14

14

14

14

2

2

2

2

2

2

14

FIJI

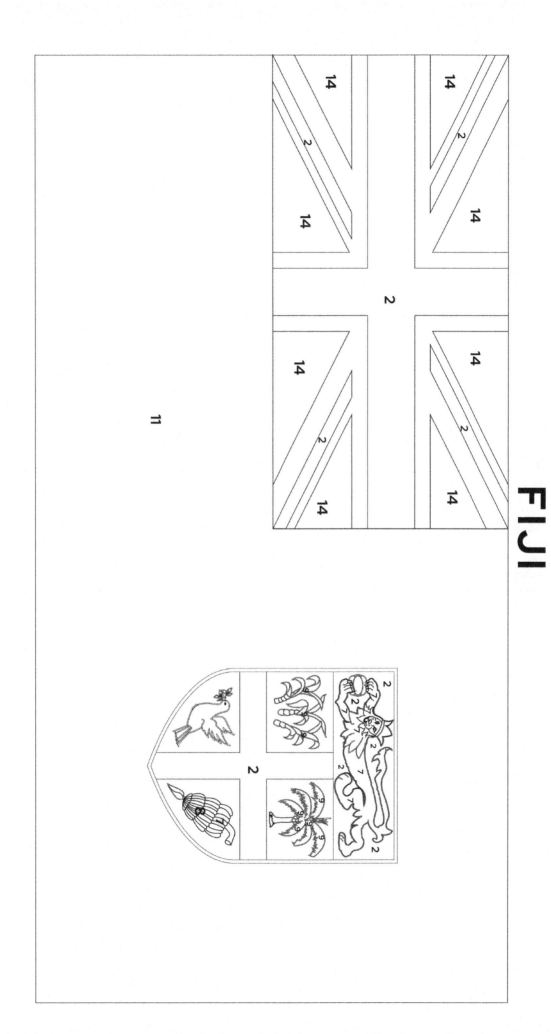

1 = Maroon
2 = Dark Red
3 = Red
4 = Orange
5 = Light Orange
6 = Golden Yellow
7 = Light Yellow
8 = Light Green
9 = Green
10 = Dark Green
11 = Light Blue
12 = Blue
13 = Dark Blue
14 = Indigo
15 = Purple
16 = Black
17 = Brown
18 = Grey

KIRIBATI

1 = Maroon   3 = Red   5 = Light Orange   7 = Light Yellow   9 = Green   11 = Light Blue   13 = Dark Blue   15 = Purple   17 = Brown
2 = Dark Red   4 = Orange   6 = Golden Yellow   8 = Light Green   10 = Dark Green   12 = Blue   14 = Indigo   16 = Black   18 = Grey

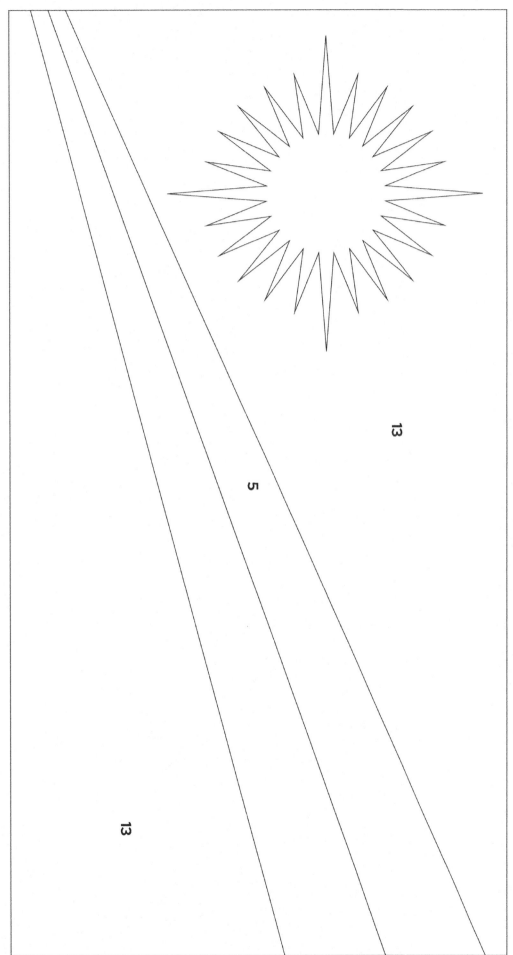

MARSHALL ISLANDS

1 = Maroon
2 = Dark Red
3 = Red
4 = Orange
5 = Light Orange
6 = Golden Yellow
7 = Light Yellow
8 = Light Green
9 = Green
10 = Dark Green
11 = Light Blue
12 = Blue
13 = Dark Blue
14 = Indigo
15 = Purple
16 = Black
17 = Brown
18 = Grey

MICRONESIA

11

NAURU

14

6

14

NEW ZEALAND

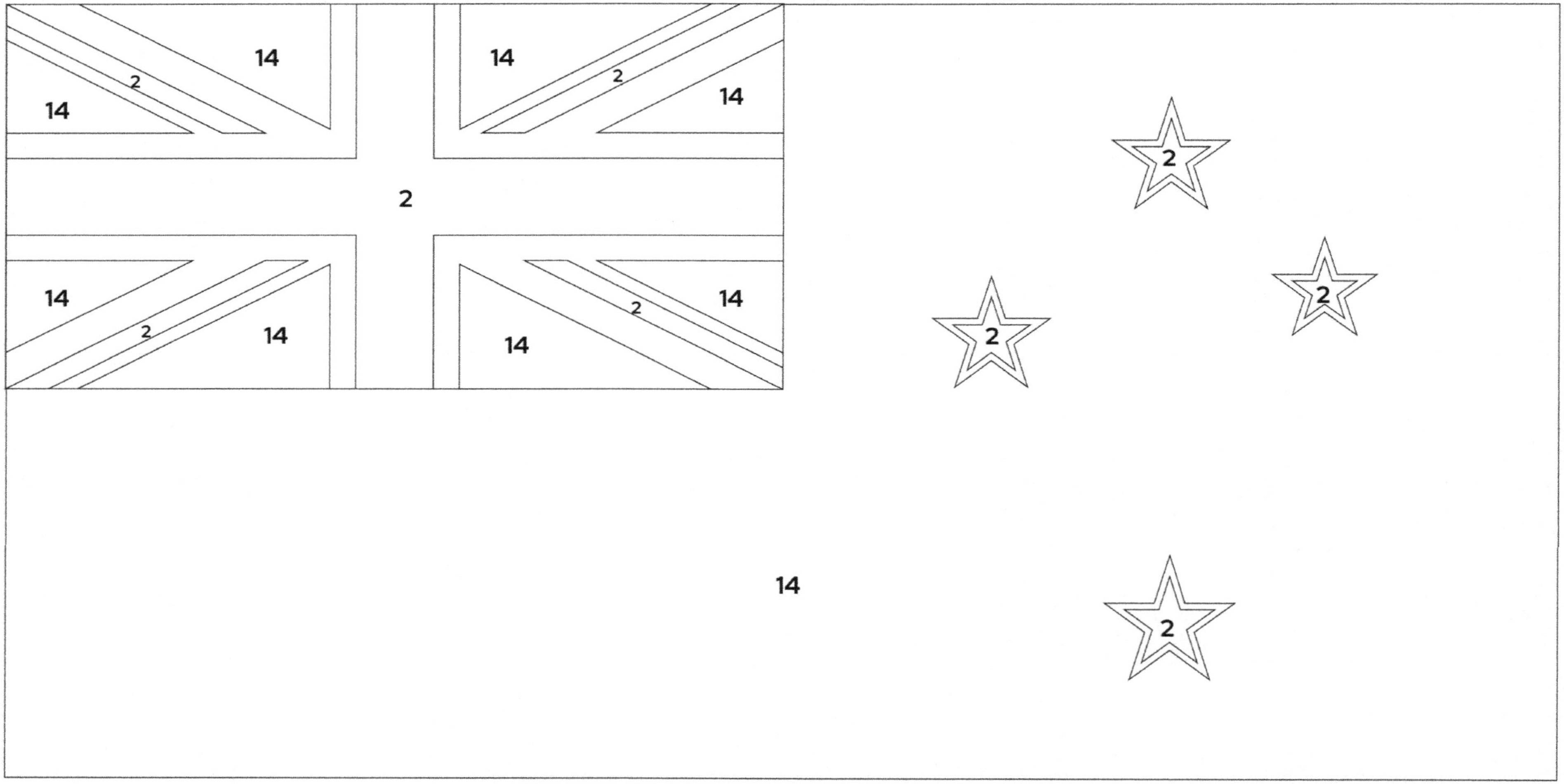

1 = Maroon
2 = Dark Red
3 = Red
4 = Orange
5 = Light Orange
6 = Golden Yellow
7 = Light Yellow
8 = Light Green
9 = Green
10 = Dark Green
11 = Light Blue
12 = Blue
13 = Dark Blue
14 = Indigo
15 = Purple
16 = Black
17 = Brown
18 = Grey

PALAU

12

7

PAPUA NEW GUINEA

1 = Maroon
2 = Dark Red
3 = Red
4 = Orange
5 = Light Orange
6 = Golden Yellow
7 = Light Yellow
8 = Light Green
9 = Green
10 = Dark Green
11 = Light Blue
12 = Blue
13 = Dark Blue
14 = Indigo
15 = Purple
16 = Black
17 = Brown
18 = Grey

SAMOA

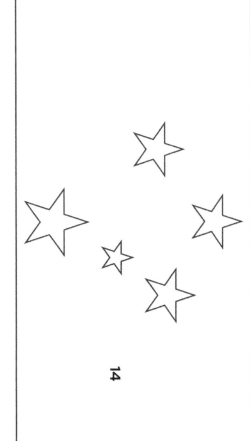

14

2

1 = Maroon
2 = Dark Red

3 = Red
4 = Orange

5 = Light Orange
6 = Golden Yellow

7 = Light Yellow
8 = Light Green

9 = Green
10 = Dark Green

11 = Light Blue
12 = Blue

13 = Dark Blue
14 = Indigo

15 = Purple
16 = Black

17 = Brown
18 = Grey

SOLOMON ISLANDS

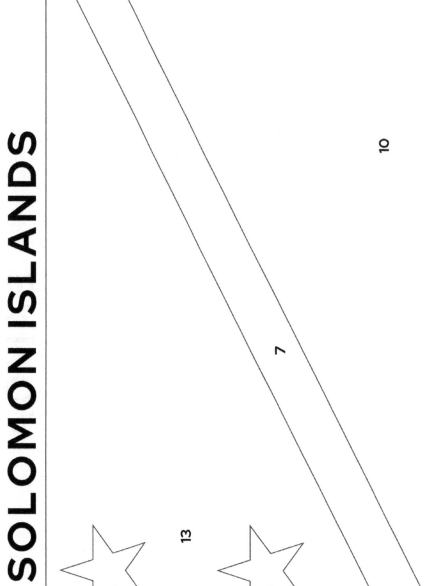

13

7

10

TONGA

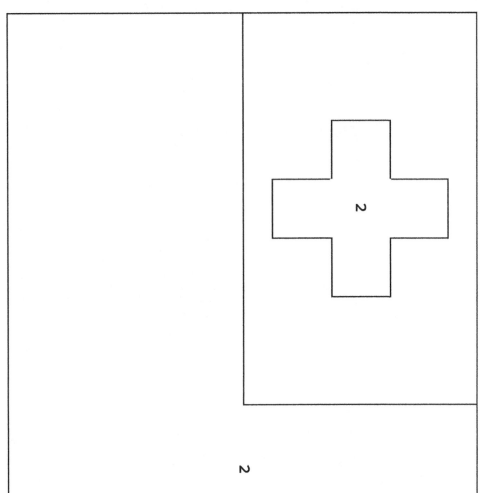

TUVALU

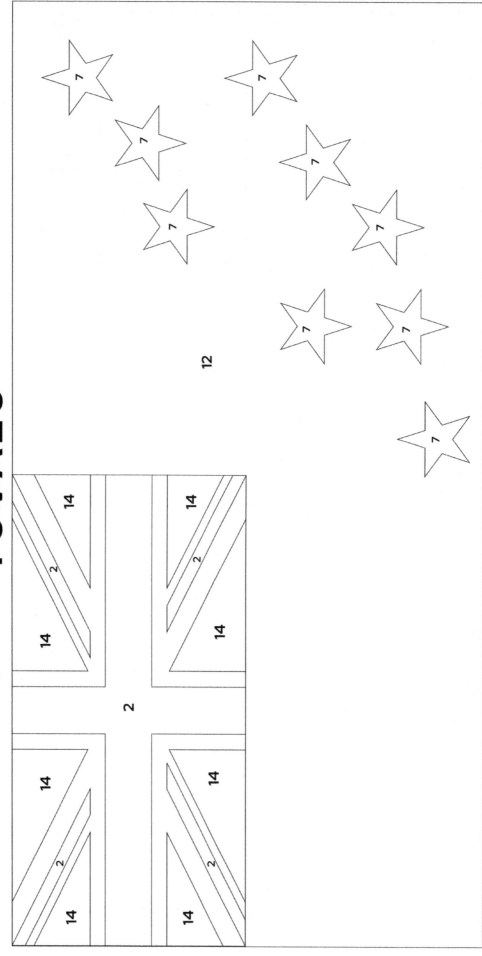

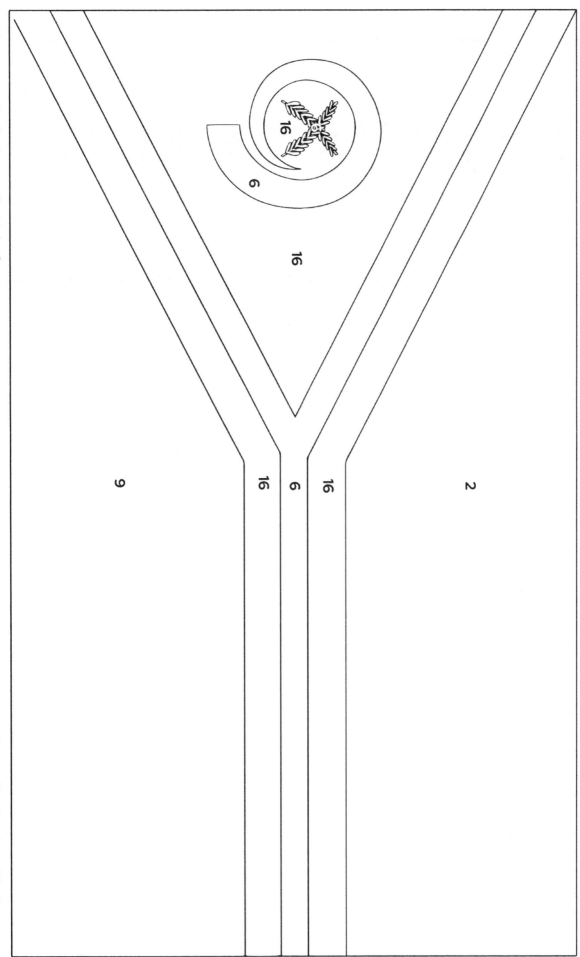

1 = Maroon
2 = Dark Red
3 = Red
4 = Orange
5 = Light Orange
6 = Golden Yellow
7 = Light Yellow
8 = Light Green
9 = Green
10 = Dark Green
11 = Light Blue
12 = Blue
13 = Dark Blue
14 = Indigo
15 = Purple
16 = Black
17 = Brown
18 = Grey

BONUS FLAGS

ENGLAND

2

1 = Maroon **3** = Red **5** = Light Orange **7** = Light Yellow **9** = Green **11** = Light Blue **13** = Dark Blue **15** = Purple **17** = Brown
2 = Dark Red **4** = Orange **6** = Golden Yellow **8** = Light Green **10** = Dark Green **12** = Blue **14** = Indigo **16** = Black **18** = Grey

EUROPEAN UNION

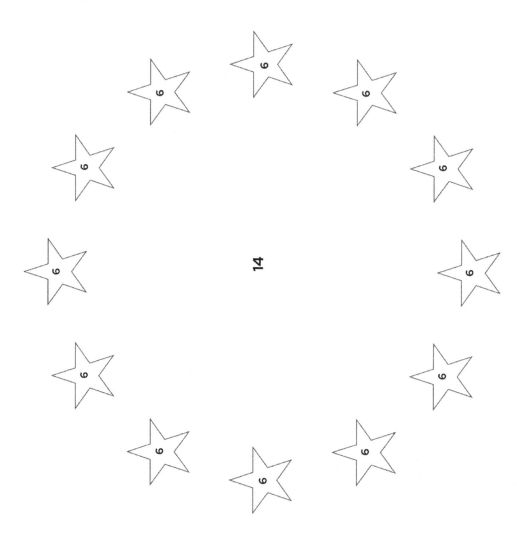

GREENLAND

3

3

HONG KONG

3

1 = Maroon **3** = Red **5** = Light Orange **7** = Light Yellow **9** = Green **11** = Light Blue **13** = Dark Blue **15** = Purple **17** = Brown
2 = Dark Red **4** = Orange **6** = Golden Yellow **8** = Light Green **10** = Dark Green **12** = Blue **14** = Indigo **16** = Black **18** = Grey

PUERTO RICO

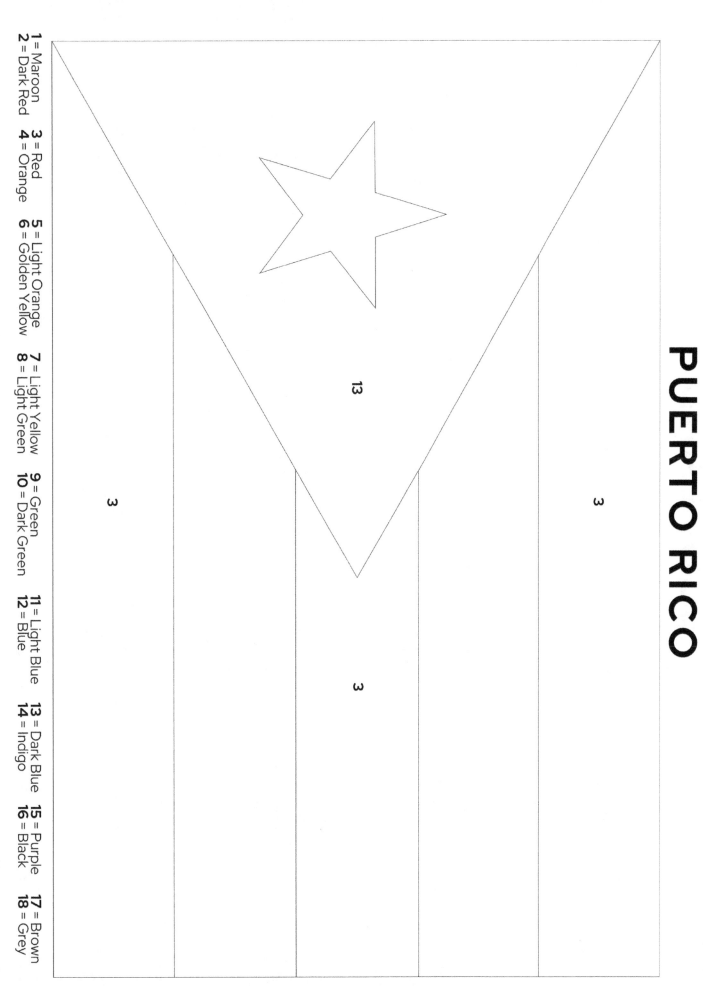

1 = Maroon
2 = Dark Red
3 = Red
4 = Orange
5 = Light Orange
6 = Golden Yellow
7 = Light Yellow
8 = Light Green
9 = Green
10 = Dark Green
11 = Light Blue
12 = Blue
13 = Dark Blue
14 = Indigo
15 = Purple
16 = Black
17 = Brown
18 = Grey

SCOTLAND

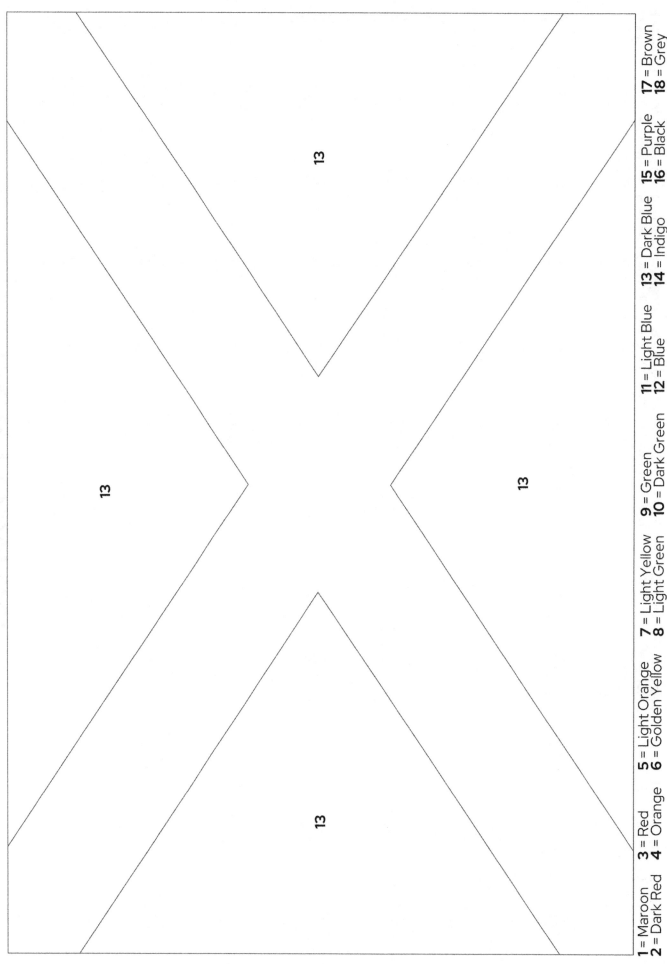

TAIWAN

14

3

1 = Maroon
2 = Dark Red

3 = Red
4 = Orange

5 = Light Orange
6 = Golden Yellow

7 = Light Yellow
8 = Light Green

9 = Green
10 = Dark Green

11 = Light Blue
12 = Blue

13 = Dark Blue
14 = Indigo

15 = Purple
16 = Black

17 = Brown
18 = Grey

WALES